STANDARD GRADE

CHEMISTRY

Iain Brand

Hodder Gibson
A MEMBER OF THE HODDER HEADLINE GROUP

The Publishers would like to thank the following for permission to reproduce copyright material:

Acknowledgements

Artworks by Peters and Zabransky.

Cartoons © Moira Munro 2005.

Every effort has been made to trace all copyright holders, but if any have been inadvertently overlooked the Publishers will be pleased to make the necessary arrangements at the first opportunity.

Orders: please contact Bookpoint Ltd, 130 Milton Park, Abingdon, Oxon OX14 4SB. Telephone: (44) 01235 827720. Fax: (44) 01235 400454. Lines are open from 9.00–5.00, Monday to Saturday, with a 24-hour message answering service. Visit our website at www.hoddereducation.co.uk. Hodder Gibson can be contacted direct on: Tel: 0141 848 1609; Fax: 0141 889 6315; email: hoddergibson@hodder.co.uk

First published in 2005 by
Hodder Gibson, a member of the Hodder Headline Group, an Hachette Livre UK company
2a Christie Street
Paisley PA1 1NB

Impression number 10 9 8 7 6 5 4 3
Year 2010 2009 2008 2007

Cover photo © Simon Lewis/Science Photo Library (TB07/036)
Typeset in 10.5 on 14pt Frutiger Light by Phoenix Photosetting, Chatham, Kent
Printed and bound in Great Britain by Martins The Printers, Berwick-upon-Tweed

A catalogue record for this title is available from the British Library

ISBN-13: 978-0-340-88550-5

CONTENTS

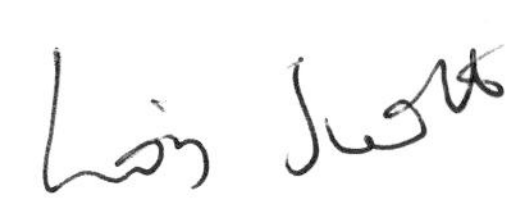

INTRODUCTION

So you want to pass Standard Grade Chemistry. Good! Chemistry is without doubt the most important science, it links well with both physics and biology, and is involved in many industries and occupations. Having chemistry as one of your qualifications will give you a great choice of careers, from medicine to oil refining and forensic science to cosmetics.

Indeed in many ways our whole lives rely on chemistry. Think about all the different substances we use to improve our lives. Things like metals, plastics, paints, fuels, cosmetics and medicines are all the result of chemical research and chemical reactions. Unfortunately, sometimes we misuse these substances, causing pollution, damaging our health and wasting resources. Serious problems like these cannot be solved unless we understand our material world. We cannot leave everything to the experts, and common sense is not always enough. Everyone needs some chemical sense.

Figure 0.1 Chemistry, a great choice for your future

An unfortunate aspect of studying chemistry, however, is the need to sit and pass examinations. Standard Grade Chemistry exams test your problem solving skills as well as your knowledge and understanding. This book has been specifically written to help you succeed in the Standard Grade Chemistry. It contains all the required knowledge and understanding, along with help on problem solving and preparation for your exams and exam techniques. This book has been designed to show you how to pass Standard Grade Chemistry and will undoubtedly be a great help to you as you prepare for your exams.

Good luck!

Iain Brand

PREPARING FOR STANDARD GRADE CHEMISTRY

Organised preparation is the key to success in all tests and examinations. This is particularly true for chemistry as it is a subject that requires a sound knowledge of the facts and theories, so that they can be applied in different situations. Remember the old motto: 'Fail to prepare, prepare to fail!' Well, you have already taken the first steps to being well prepared; you have got hold of this book. So what else do you need to do?

Getting organised

When preparing for your Standard Grade Chemistry, you will need to be well organised, with your books, notes and other learning resources in order and easily available.

You should be able to find some or all of the following:

- this book, *How to Pass Standard Grade Chemistry*
- class notes and summary notes
- past paper questions and Data Booklet
- blank paper, pencils, pens, highlighters, a ruler and a calculator.

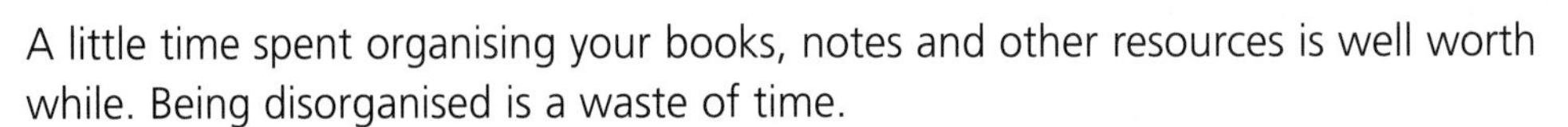

A little time spent organising your books, notes and other resources is well worth while. Being disorganised is a waste of time.

Planning how to study

Do you enjoy 'swotting' for exams? Few people do, so even after you have organised your notes and books, making a start is often difficult. What you need is a 'study timetable' to organise *what*, *when* and even *how* you are going to study. It should include all of your subjects, not just chemistry, and it is best to vary your methods of study as doing the same thing for hours can be inefficient. Your

Figure 0.2 Enjoying studying?

timetable should therefore divide the day into 30 minutes intervals and set up different activities for each period.

It is also useful to have a subject 'study plan' which lists the topics to be studied and allows you to record the progress of your revision.

An example of one Standard Grade Chemistry study plan can be found in Appendix I.

It is also a good idea to award yourself a number of 'stars' depending on how well you know a topic. This will clearly show where extra revision is required.

Improving knowledge and understanding

Start your revision with an easy topic, as this builds confidence. In Standard Grade Chemistry the units of work on 'Chemical reactions' and 'Reaction Rates' are an obvious starting point. Try to vary your revision activities: you can't just read your notes and expect the information to 'stick' in your brain. You need to help your memory and understanding by taking notes, making summaries, listing key words, etc. Your notes and summaries don't need to be neat, but they should be in your own words.

Practising exam techniques

The final piece in the preparation 'jigsaw' is practising your **exam technique**. You must become familiar with the style of the exam paper and the type of question asked if you are to show off your knowledge and skills to their best. Practising **past paper** questions is very important.

Figure 0.3 Do you know how to succeed in exams?

Problem solving

Problem solving is one of the two main elements in Standard Grade Chemistry and makes up about half the questions in all examinations.

Problem solving tips

- Always read the question carefully and note the meanings of **headings**, **keywords**, **scales** and **units**.
- To describe how to carry out an investigation, draw a **labelled** outline **diagram**, list the steps involved and explain how the test is made **fair**.
- If required name and describe observations, measuring instruments and **units** used.
- When obtaining information from a line graph use a ruler and note the scales and units carefully.
- Note the type of graph required. Pupils often mix up bar graphs, and line graphs
- When drawing a graph choose a scale to take up more than half the graph paper and write the names and units on each axis.
- Draw the '**best fit**' curve or straight line in a line graph. This line may not go through all the points.
- To predict a value using a line graph extend the line to include other values.
- When asked how two variables are related describe the direction of the relationship. You could word your answer in the form, 'As *x* increases *y* . . . . . .'
- When describing a chemical test state what you do and what happens. If a colour change occurs, describe the colour at the start and at the end. You should know the tests for: *oxygen, carbon dioxide, hydrogen, alkenes, glucose, starch and* Fe^{2+} *ions*.
- Show **all your working** in calculations and remember to use the correct **units** (g, cm^3, moles etc.).
- Get to know your data book. It contains lots of useful information about bonding, formulae, equations etc.

Chapter 1

INTRODUCTION TO CHEMISTRY AND CHEMICAL REACTIONS

So what is chemistry? Is it, as some people think, just about dangerous acids, strange coloured liquids and funny smells? Of course not; although these things have a part to play in chemistry, there is much more to the subject. This topic tries to explain what chemistry and chemical reactions are all about.

Key Words

★ **atom** ★ **chemical equation**
★ **chemical reaction**
★ **compound** ★ **effervescence**
★ **element** ★ **exothermic reaction** ★ **filtration**
★ **molecule** ★ **Periodic Table**
★ **precipitation** ★ **products**
★ **properties** ★ **reactants**
★ **word equation**

Figure 1.1 Is this chemistry?

Matter, atoms and molecules

In simple terms, chemistry is the study of the structure and properties of **matter**. Through the study of chemistry you should gain a better understanding of all the substances that make up the material world.

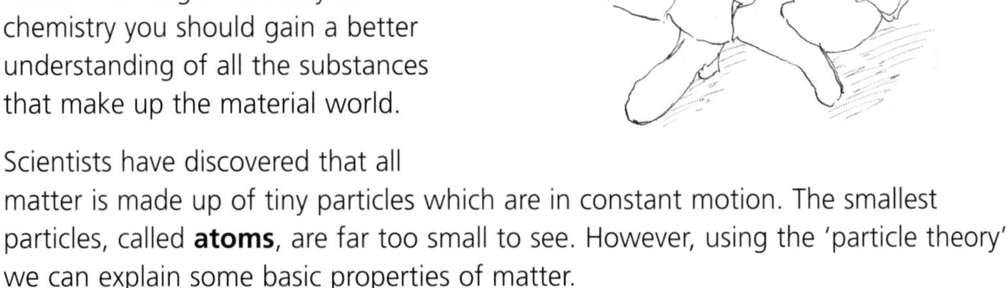

Scientists have discovered that all matter is made up of tiny particles which are in constant motion. The smallest particles, called **atoms**, are far too small to see. However, using the 'particle theory' we can explain some basic properties of matter.

In some substances the atoms are joined together in small groups called **molecules**.

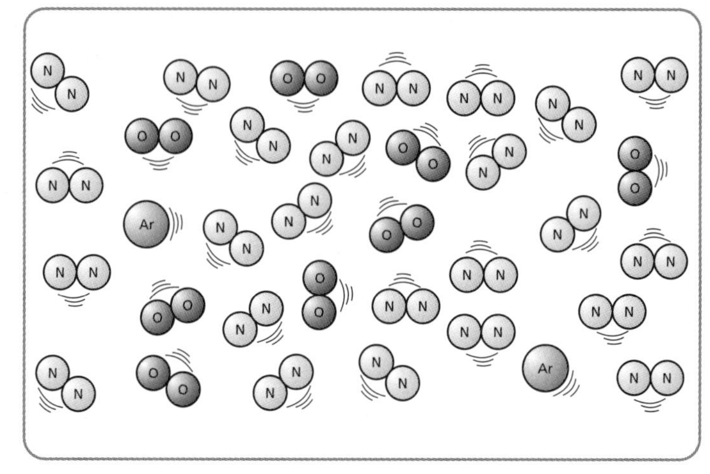

Figure 1.2 Air contains a mixture of different molecules

Chemical reactions

To understand chemistry you must first understand the changes that form new substances. These changes are called **chemical reactions**, and they occur all around us. Fuels burning, plants growing, cars rusting, foods cooking and the manufacture of all sorts of substances from metals to medicines are all examples of chemical reactions. A chemical reaction always forms one or more new substances.

Figure 1.3 Chemical reactions happen all around us

We also carry out many chemical reactions in the laboratory. During these reactions you can sometimes see the new substances being formed by:

- A change in **colour**.
- **Effervescence**, a gas being produced from a solution.
- **Precipitation**, a solid being produced from solutions.

Many chemical reaction also involve an energy change. Usually heat energy is given out and sometimes light energy as well. A reaction that gives out heat is called an **exothermic reaction**.

Note that changes of state like melting, freezing, evaporation and condensation are physical changes. New substances are *not* formed, so they are *not* chemical reactions.

Figure 1.4 Combustion is an exothermic reaction

The substances you start with in a chemical reaction are called the **reactants** and the new substances that are formed are called the **products**. Chemists show these changes by writing **chemical equations**. In equations the reactants are written on the left and the products are written on the right. An arrow in between shows the direction of change.

For example, during photosynthesis plants take in carbon dioxide and water and make glucose, an important food, and oxygen, the gas essential to life. The **word equation** for this reaction is:

$$\text{carbon dioxide} + \text{water} \rightarrow \text{glucose} + \text{oxygen}$$
$$\textbf{reactants} \rightarrow \textbf{products}$$

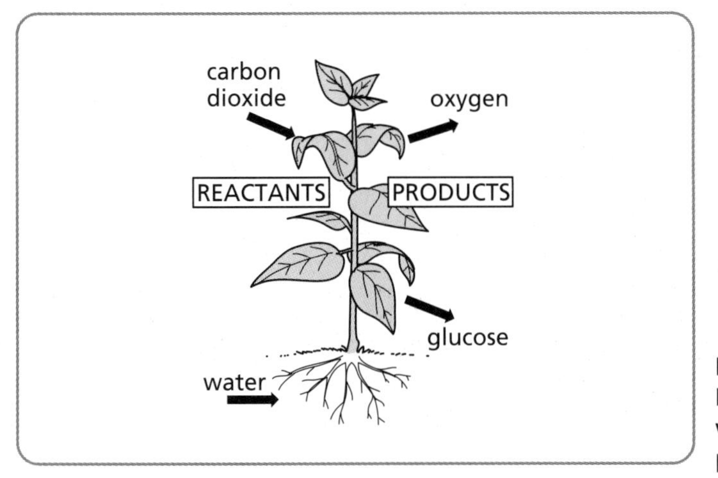

Figure 1.5 Photosynthesis, a reaction which occurs in green plants

Key point

In a chemical equation: **REACTANTS → PRODUCTS**

Elements and the Periodic Table

Elements are the simplest substances in nature. They cannot be broken down or made simpler because each element is made up of only one kind of atom. For example the element iron contains only iron atoms. If you heat it, cool it or, break it up it still only contains iron atoms.

All the matter in the universe is made up of about 100 elements. Each of these elements has a **symbol** and an **atomic number**. The names, symbols and atomic numbers of all the elements can be found in a special list called the **Periodic Table**.

In the **Periodic Table** the elements are placed in order of their atomic number into vertical columns called **groups**. The Periodic Table is useful as it gives you information about the properties of the elements. For example, **metals** are found on the left side of the table and **non-metals** on the right. The Periodic Table of the elements is dealt with in more detail in Chapter 3.

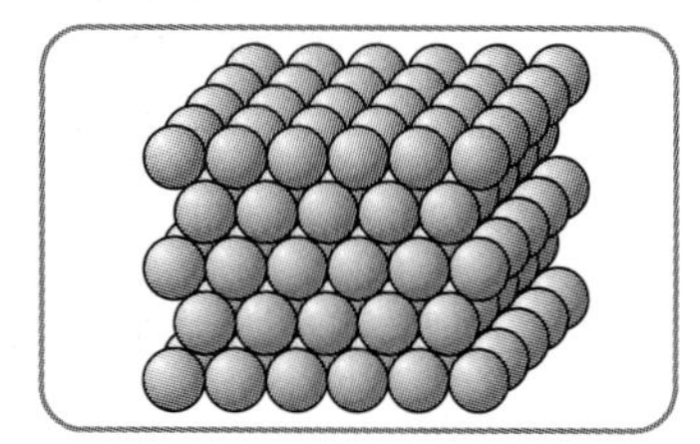

Figure 1.12 Atoms of the element iron

I	II											III	IV	V	VI	VII	VIII
									1 **H** hydrogen								2 **He** helium
3 **Li** lithium	4 **Be** beryllium											5 **B** boron	6 **C** carbon	7 **N** nitrogen	8 **O** oxygen	9 **F** flourine	10 **Ne** neon
11 **Na** sodium	12 **Mg** magnesium											13 **Al** aluminium	14 **Si** silicon	15 **P** phosphorus	16 **S** sulphur	17 **Cl** chlorine	18 **Ar** argon
19 **K** potassium	20 **Ca** calcium	21 **Sc** scandium	22 **Ti** titanium	23 **V** vanadium	24 **Cr** chromium	25 **Mn** manganese	26 **Fe** iron	27 **Co** cobalt	28 **Ni** nickel	29 **Cu** copper	30 **Zn** zinc	31 **Ga** gallium	32 **Ge** germanium	33 **As** arsenic	34 **Se** selenium	35 **Br** bromine	36 **Kr** krypton
37 **Rb** rubidium	38 **Sr** strontium	39 **Y** yttrium	40 **Zr** zirconium	41 **Nb** niobium	42 **Mo** molybdenum	43 **Tc** technetium	44 **Ru** ruthenium	45 **Rh** rhodium	46 **Pd** palladium	47 **Ag** silver	48 **Cd** cadmium	49 **In** indium	50 **Sn** tin	51 **Sb** antimony	52 **Te** tellurium	53 **I** iodine	54 **Xe** xenon
55 **Cs** caesium	56 **Ba** barium		72 **Hf** hafnium	73 **Ta** tantalum	74 **W** tungsten	75 **Re** rhenium	76 **Os** osmium	77 **Ir** iridium	78 **Pt** platinum	79 **Au** gold	80 **Hg** mercury	81 **Tl** thallium	82 **Pb** lead	83 **Bi** bismuth	84 **Po** polonium	85 **At** astatine	86 **Rn** radon

Figure 1.6 The Periodic Table of the elements

Compounds and mixtures

When mixtures of elements react their atoms join together to form a **compound**. Therefore a compound contains two or more elements joined together. It is usually more difficult to separate the elements in a compound than the elements in a mixture.

The diagram below shows how the elements hydrogen and chlorine form the compound hydrogen chloride.

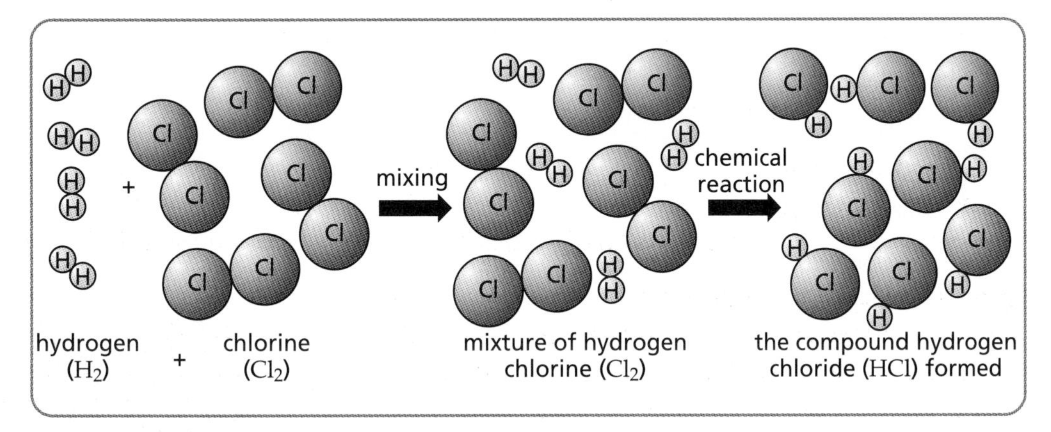

Figure 1.7 Making a compound

Although there are only about 100 elements, chemists have identified millions of different compounds.

The name of a compound sometimes tells you the names of the elements it contains. For example, in two-element compounds the ending of one of the names is changed to –ide. If the ending is changed to –ate or –ite the compound contains oxygen as well as two other elements. Some examples of the names of compounds are shown in the table below.

Name of compound	Elements joined together
sodium chloride	sodium and chlorine
zinc oxide	zinc and oxygen
lead sulphide	lead and sulphur
copper sulphate	copper, sulphur and oxygen
iron nitrite	iron, nitrogen and oxygen

Properties of matter

The **properties** of a substance describe what it looks like or what it does. Understanding the properties of substances is important, in choosing a material for a particular job. Properties are generally grouped under two headings, as shown below.

Key Points

Chemical properties describe what a substance does in chemical reactions.	Physical properties describe the appearance of a substance or what it does without reacting.
◆ flammability (how easily a substance catches fire)	◆ colour and state
◆ reaction with acid	◆ strength and hardness
◆ solubility in water	◆ melting point and boiling point
◆ pH of solution	◆ electrical conductivity and thermal conductivity

Summary

- During a chemical reaction one or more new substances are always formed.
- Examples of everyday reactions include rusting, cooking foods and striking a match.
- A chemical reaction can be identified by a change in appearance or energy.
- A precipitation reaction forms a solid, an effervescence reaction produces a gas.
- A reaction which gives out heat is called an exothermic reaction.
- Mixing and changes in state (evaporation, freezing, etc.) are *not* chemical reactions.
- Atoms and molecules are the small particles which make up everything.
- An element contains only one kind of atom; there are about 100 elements.
- The elements are listed in the Periodic Table, metals on left and non-metals on right.
- Compounds are formed when elements react together.
- A compound contains the atoms of two or more elements joined together.
- ...ide compounds contain two elements, ...ate or ...ite also contain oxygen.

Chapter 2

SPEED OF REACTIONS

You know that chemical reactions form new substances, but how quickly do reactions happen? Do they all occur at the same speed and can you change the speed of a reaction? This chapter looks at ways of measuring the rate of reactions and how varying the conditions can change the rate.

Key Words

* **catalyst** * **concentration** * **products** * **reactants** * **reaction rate**
* **variable**

How chemical reactions occur

The **rate** of a chemical reaction can vary greatly from the very slow, as in the rusting of iron, to the almost instantaneous explosion or precipitation reaction.

Chemical reactions are not some sort of 'magic'. They simply involve rearranging atoms and molecules to form new substances. Understanding how chemical reactions occur will help you explain how they can be controlled.

Scientists have found that reaction only takes place when the reactant molecules **collide** with enough **energy** or speed. During these collisions the atoms and molecules are rearranged to form new substances, the products of the reaction.

Changing reaction rates

Sometimes it is necessary to control the speed of a chemical reaction, making it faster or slower. The main factors which control the rate of reaction are **concentration**, **temperature** and **particle size**. These are called **variables**, and they can be investigated by carrying out experiments like the ones described on the following page.

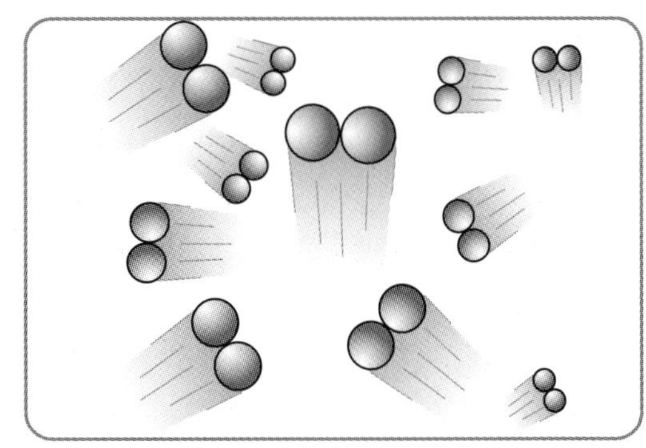

Figure 2.1 Reactant molecules collide to form the products

Investigating concentration

In reactions involving solutions the reaction is faster the higher the concentration of the solution. For example, magnesium reacts faster with a more concentrated acid.

The concentration of a solution is measured in moles per litre (mol/l).

Investigating temperature

Increasing the temperature will increase the rate of a reaction. For example, zinc reacts faster with hydrochloric acid when the acid is heated.

Investigating particle size

In reactions involving solids, the reaction is faster the smaller the pieces of solid used. For example, the smaller the marble chips the faster the reaction with acid.

The smaller pieces of solid have a larger surface area, so there are more collisions between the reactant molecules and hence a faster reaction.

Figure 2.2 Out of control?

Key Points

A summary of how changing conditions affect reaction rates is give in the table below. Some everyday examples of changing reaction rates are also included in the table.

Variable change	Effect on reaction rate	Everyday example
Increasing concentration	Increasing	More concentrated bleach cleans faster than dilute bleach.
Increasing temperature	Increasing	Food cooks faster if the temperatures of the oven is increased.
Decreasing particle size	Increasing	Wood chips burn faster on a fire than lumps of wood.

Catalysts

Another way of changing the speed of a reaction is to use a **catalyst**. These are substances that speed up a chemical reaction without being used up. At the end of the reaction the catalyst can be recovered unchanged.

For example, a platinum catalyst is used in the 'catalytic converter' in car exhausts to change poisonous gases into harmless gases.

The platinum catalyst is spread over a honeycomb structure to give it a large surface area.

Catalysts are also very important in the chemical industry. They save energy and make manufacturing processes more economic. For example, **iron** is used in the **Haber process** to make ammonia.

Measuring reaction rates

To study reaction rates you need to measure the changes taking place during reactions. For example, in reactions that produce a gas you can measure the volume of gas given off, or the loss in mass, with time. The set-up in Figure 2.3 could be used to investigate how the size of chips affects the rate of reaction between marble and hydrochloric acid.

To make a test fair only one variable can be changed at a time; the other variables have to be controlled. In this particular experiment the particle size was changed. Therefore the type and concentration of the acid, the temperature and the mass of marble had to be kept the same.

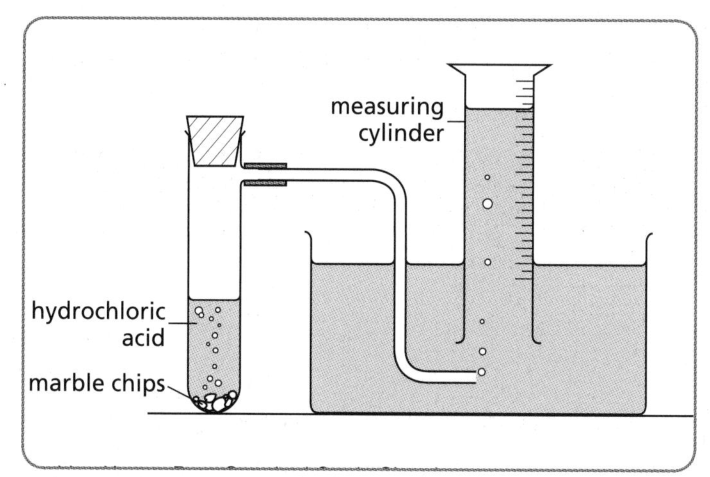

Figure 2.3 Measuring rates of reaction

During the reaction the volume of gas was measured at regular intervals and a graph of volume of gas against time was drawn. This graph shows what happened to the speed of the reactions.

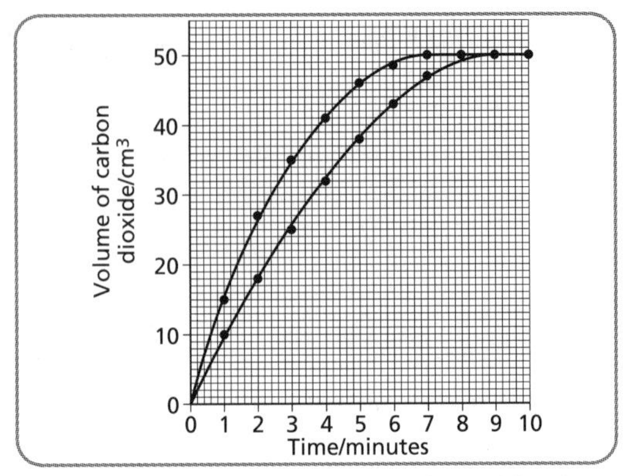

Figure 2.4 Comparing sizes of marble chips

The results showed that the reaction with the smaller chips was faster, as it finished in a shorter time. It also showed that reactions are faster at the start and get slower with time. This is because the concentrations decrease as the reaction proceeds. Eventually the reactants are used up and the reaction stops.

Summary

- Reactions can vary in speed, from slow (like rusting) to fast (like explosions).
- Increasing concentration (higher mol/l) increases the speed of a reaction.
- Increasing temperature (heating) increases the speed of a reaction.
- Decreasing particle size (using powders) increases the speed of a reaction.
- Catalysts speed up reactions without being used up, e.g. platinum in catalytic converters.
- You can follow the rate of a reaction by measuring volumes of gas formed or mass lost.

Chapter 3

ATOMS AND THE PERIODIC TABLE

What are atoms made up of? How is one element different from another and why is the Periodic Table so important? This chapter introduces atomic structure and explains the arrangement of the elements in the Periodic Table.

Key Words

★ alkali metal ★ atom ★ atomic number ★ electron ★ electron energy level/shell ★ halogen ★ ion ★ isotope ★ mass number ★ monatomic ★ neutron ★ noble gas ★ nucleus ★ Periodic Table ★ proton ★ relative atomic mass ★ transition metal

Classifying elements

The materials of our world are made from about 100 different elements and millions of different compounds. To make sense of this variety of materials chemists like to classify them into groups according to their properties. There are several ways that elements can be classified:

- Solids, liquids or gases. At room temperature most elements are solids, there are eleven gases and only two liquids.
- Metals or non-metals. About three-quarters of the known elements are metals and the rest are non-metals.

By far the most useful classification of elements is the **Periodic Table**. This groups the elements by arranging them in increasing **atomic number**, in **groups** (vertical columns) and **periods** (horizontal rows).

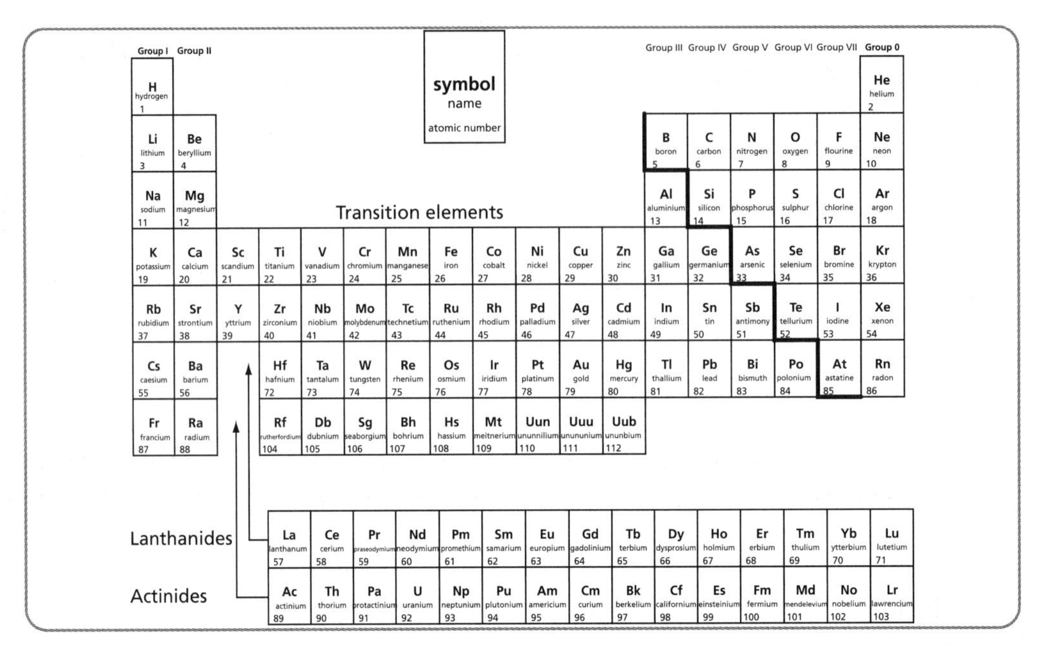

Figure 3.1 In the modern Periodic Table the elements are arranged in order of their atomic number

The properties of elements

The Periodic Table was arranged so that elements in the same group had similar chemical properties.

Example

Examples of the similarities of some of the main groups are shown below.

Group 1 – The alkali metals

Li
Na
K
Rb
Cs

All the elements in this group:

- Are metals which are shiny when cut but dull quickly in air.
- React quickly with cold water forming an alkaline solution and hydrogen.
- Are stored under oil.

Group 7 – The halogens

F
Cl
Br
I

All the elements in this group:

- Are non-metals.
- Exist as diatomic molecules.
- React quickly with metals.

***Example** continued* ➢

Example continued

Group 0 – The noble gases

He All the elements in this group:
Ne ◆ Are non-metal gases.
Ar ◆ Exist as single atoms (**monatomic**)
Kr ◆ Are very unreactive.
Xe

The elements in the central block of the Periodic Table are called **transition metals**. Although some react faster than others, they have many similar chemical properties.

Atoms

All matter is made up of tiny particles called **atoms**. So to understand the properties of matter you need to know about the structure of atoms. Scientists have found that all atoms are made up of three smaller particles called **protons**, **neutrons** and **electrons**.

The three sub-atomic particles have different properties of mass and charge.

Particle	Charge	Relative mass
proton	positive	1
electron	negative	$\frac{1}{1850}$ (negligible)
neutron	no charge	1

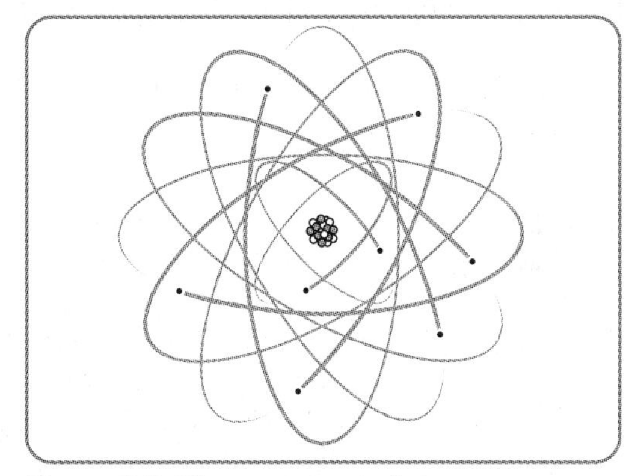

Figure 3.2 The structure of an atom

The mass of these particles is so small that we do not measure them in grams. We say that the protons and neutrons have a relative mass of 1. The mass of an electron is very much smaller. Sometimes we ignore it and say that electrons have a *negligible* mass.

Atoms of different elements have different numbers of protons, neutrons and electrons. However, all atoms have a tiny **nucleus**, surrounded by even smaller electrons. Most of the mass of any atom is in its nucleus, while the rest is mainly empty space.

Describing an atom

Two numbers are used to describe any atom.

- The **atomic number** is the number of protons in the nucleus. Atoms of the same element always have the same number of protons and so the same atomic number.
- The **mass number** is the total number of protons and neutrons in the nucleus. Atoms of the same element can have different mass numbers.

All atoms are *neutral*. They must have the *same number of protons and electrons*, so that the opposite charges cancel out. Therefore given the atomic number and mass number of an atom, you can work out its structure in terms of protons, neutrons and electrons.

Key Point

protons = atomic number

neutrons = mass number – atomic number

electrons = protons

For example, lithium with an atomic number of 3 and a mass number of 7, has 3 protons, 4 neutrons and 3 electrons

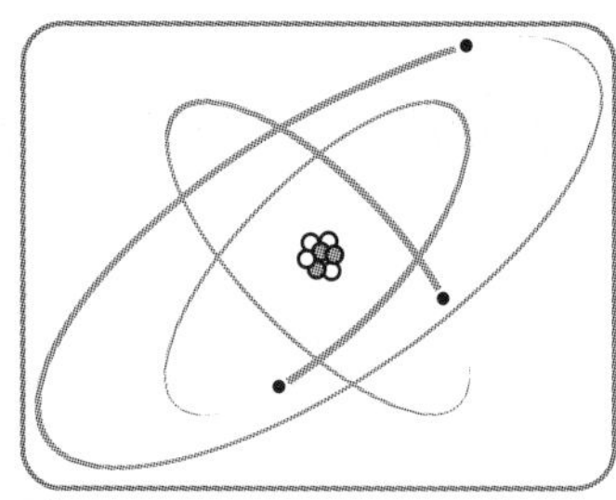

Figure 3.5 An atom of lithium

The atomic number and mass number of an atom are often given in the form shown below.

Key Point

$$^{\text{mass no.}}_{\text{atomic no.}}\text{Symbol}$$

The table below shows some more examples of atomic structures.

Element	Atom	Atomic no.	Mass no.	Numbers of . . .		
				protons	neutrons	electrons
Hydrogen	$^{1}_{1}H$	1	1	1	0	1
Iron	$^{56}_{26}Fe$	26	56	26	30	26
Iodine	$^{127}_{53}I$	53	127	53	74	53
Uranium	$^{235}_{92}U$	92	235	92	143	92

Atoms and ions

In certain circumstances an atom can gain or lose electrons. If this happens it is no longer neutral, but has becomes a charged particle called an **ion**. Positive ions are formed by the loss of electrons and negative ions are formed by the gain of electrons. The charge on the ion depends on the number of electrons lost or gained.

For example, a lithium atom, $^{7}_{3}Li$ (3p, 4n and 3e) can lose 1 electron to form a single positive lithium ion, $^{7}_{3}Li^{+}$ (3p, 4n and 2e).

The table below shows some other examples of ion structures.

Ion	Atomic no.	Mass no.	Numbers of . . . protons	neutrons	electrons
$^{27}_{13}Al^{3+}$	13	27	13	14	10
$^{32}_{16}S^{2-}$	16	32	16	16	18
$^{81}_{35}Br^{1-}$	35	81	35	46	36

The charge on the ion is always written at the top right hand side of the symbol. The formation of ions will be dealt with in more detail in Chapter 7.

Electron arrangements

A reaction between elements occurs when their atoms collide. As it is the electrons that meet and interact during these collisions, the properties of an element must depend on the **electron arrangement** of its atoms.

Scientists have found that the electrons in all atoms are arranged in areas called **electron shells** or **energy levels**. Each shell can hold only a certain number of electrons. The electron arrangements of some elements are shown on the next page. In these 'target diagrams' each circle represents an electron shell.

These target diagrams don't look like the real atoms. They just try to show how the electrons are arranged in the shells. The diagrams would be difficult to draw for larger atoms. So we usually just write the electron arrangements in numbers, and these can be found on page 1 of your *Chemistry Data Booklet*.

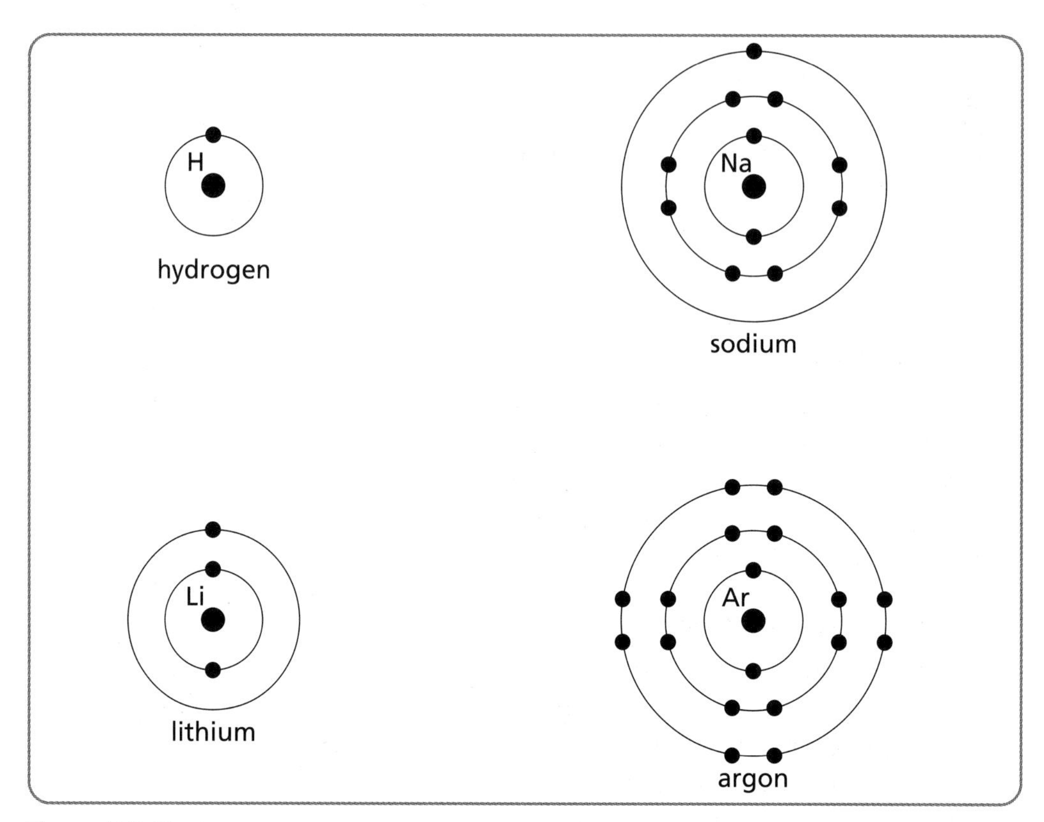

Figure 3.3 Electron arrangements

Electron arrangements and the Periodic Table

The organisation of the elements into groups in the Periodic Table was only really understood when the electron arrangements of the elements were discovered.

It seems that elements in the same group have similar chemical properties because they have the same number of electrons in their outer shell. Therefore the way an element reacts depends on its electron arrangement and in particular on the number of electrons in the outer shell of its atoms.

Isotopes and relative atomic mass

Atoms of the same element must have the same atomic number but can have different mass numbers. Atoms like this are called **isotopes**. They have the same number of protons and electrons but different numbers of neutrons.

For example, lithium has two isotopes: $^{6}_{3}\text{Li}$ which has 3p, 3n and 3e and $^{7}_{3}\text{Li}$ which has 3p, 4n and 3e.

As atoms that are isotopes have the same numbers of protons and electrons they will react in the same way. So isotopes are just atoms of the same element which have a slightly different mass.

Relative atomic mass

The **relative atomic mass** of an element can be used in calculations involving amounts of chemicals. The relative atomic mass is the average mass of an element's isotopes and so it is rarely a whole number. The values for the relative atomic masses of selected elements can be found on page 4 of your *Chemistry Data Booklet*, where they have been rounded to the nearest 0.5.

Example

Examples of relative atomic masses

Element	Relative atomic mass
carbon	12
copper	63.5
nitrogen	14
oxygen	16
tin	118.5

Summary

- The Periodic Table arranges the elements in order of increasing atomic number in groups (↕) and periods (↔).
- Elements in same group have similar chemical properties.
- Group 1, the alkali metals, all form an alkali and hydrogen gas with water.
- Group 7, the halogens, are all reactive non-metals.
- Group 0, the noble gases, are all very unreactive gases.
- Atoms contain protons and neutrons in the nucleus with electrons around it.
-

	Proton	Neutron	Electron
Charge	positive	neutral	negative
Relative mass	1	1	0

- The atomic no. = protons. The mass no. = protons + neutrons.
- Atoms are neutral as protons = electrons, so charges cancel out.
- Atoms of the same element have the same atomic number so the same number of protons and electrons.
- The electrons are arranged in shells or energy levels.
- Elements in the same group have the same number of outer electrons.
- Ions have either lost electrons (positive ion) or gained electrons (negative ion).
- Ions are written as: ${}^{x}_{y}W^{z}$, where x is the mass number, y is the atomic number and z is the ion charge.
- Isotopes are atoms of same element with the same atomic number but different mass numbers.
- The relative atomic mass of an element is the average relative mass of its isotopes.

Chapter 4

CHEMICAL FORMULAS AND HOW ATOMS COMBINE

Most people know that H_2O is the formula of a water molecule. But how can we work out the formula of substances like water and what is it that holds the water molecules together? This topic looks at how bonds are formed between atoms in molecules and explains how to work out the chemical formulas for substances.

Key Words

★ **bond** ★ **covalent bond** ★ **chemical formula** ★ **diatomic**
★ **electrostatic** ★ **molecular formula** ★ **molecule**
★ **stable electron arrangement** ★ **state symbol** ★ **valency**

Molecules and covalent bonds

Some elements and compounds are made up of **molecules.** That is, they exist as small groups of atoms held together by **bonds**.

The bonds that hold the atoms together in molecules are called **covalent bonds**. They are usually found in non-metal elements and compounds, which are said to have a covalent **molecular structure**.

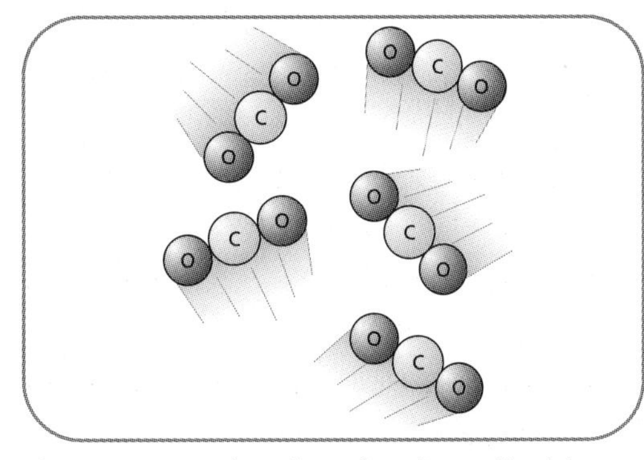

Figure 4.1 Molecules of carbon dioxide

The covalent bonds in molecules are formed by *sharing pairs of electrons* between the atoms. This happens when the outer shells of the atoms overlap. By sharing electrons the atoms get a **stable electron arrangement** like the noble gases. That is, they have 2 or 8 electrons in their outer shell.

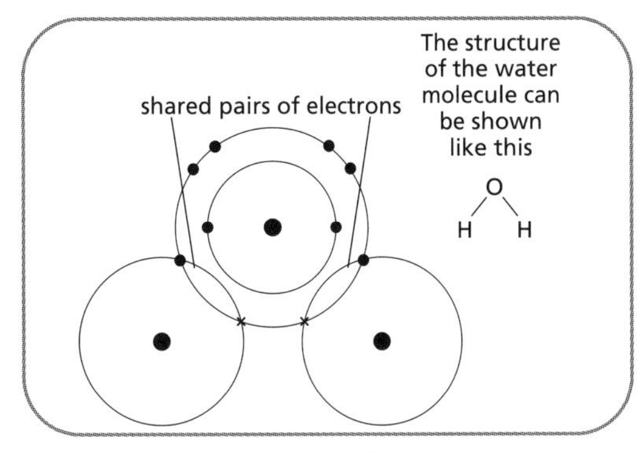

Figure 4.2 A molecule of water

You can show how atoms share electrons and form bonds by using the target diagrams for electron arrangements.

A bond is a force of attraction that holds atoms together. **Covalent bonds,** arise from the attractions between the positive nuclei and the shared electrons which are negative. Attractions like these, between positive and negative charges, are called **electrostatic** attractions. As they are strong forces of attraction the covalent bonds are hard to break.

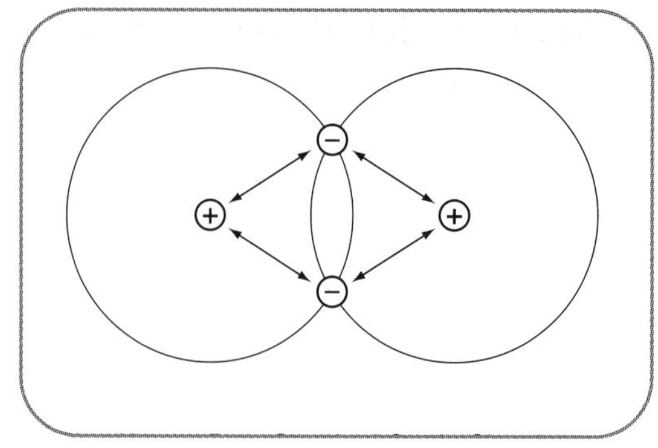

Figure 4.3 Attraction between positive nuclei and shared negative electrons holds the atoms of hydrogen together

The formulas of elements and compounds

The formula of a substance is a shorthand way of describing what it is made of. In particular, the **chemical formula** of a compound tells you the ratio of atoms of each element present. The **molecular formula** of a substance tells you the number of atoms of each element in its molecules.

The formulas of elements

The formula of a pure element is usually just its symbol. For example:

iron – Fe ; argon – Ar ; magnesium – Mg ; sulphur – S ; and so on.

However, some elements exist as **diatomic** molecules. These have two atoms bonded together in each molecule. Their formulas are written to show that they are diatomic.

Remember

There are seven diatomic elements. You can remember them with this sentence:

I_2 **I**

Br_2 **B**ring

Cl_2 **C**lay

F_2 **F**or

O_2 **O**ur

N_2 **N**ew

H_2 **H**ouse

The formulas of compounds

The formulas of some covalent compounds can be worked out from the structure of their molecules. The formula is calculated by just counting up the atoms.

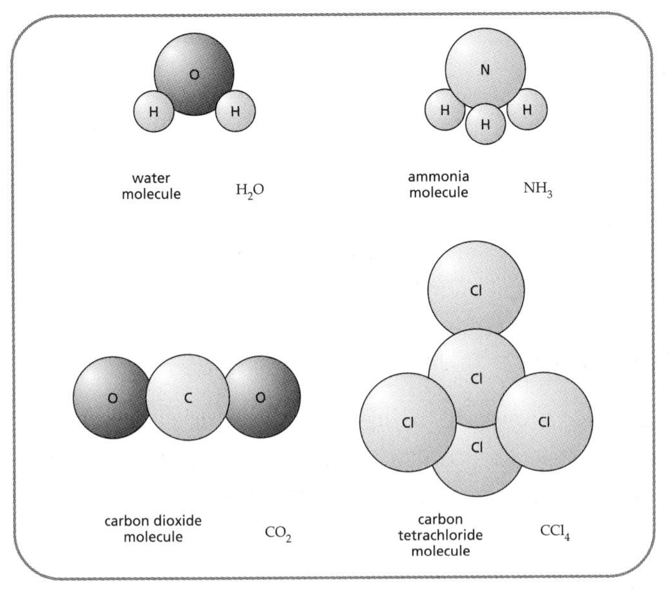

Figure 4.4 Molecules of some covalent compounds

Key Point

The molecules that are drawn above have specific structural shapes. The shape of a molecule depends on a number of factors, including the repulsion between electrons in bonds. However, full explanations are difficult so you should just remember the shape of H_2O, CO_2, NH_3, CCl_4 and related molecules.

The formulas of compounds can also be worked out. To do this you need to use the **valency** of an element. This can be described as the combining power or the number of bonds formed.

The valencies of the main group elements depend on their position in the Periodic Table.

Group 1	Group 2	Group 3	Group 4	Group 5	Group 6	Group 7	Group 8
H							He
Li	Be	B	C	N	O	F	Ne
Na	Mg	Al	Si	P	S	Cl	Ar
K	Ca	Ga	Ge	As	Se	Br	Kr
Valency 1	Valency 2	Valency 3	Valency 4	Valency 3	Valency 2	Valency 1	Valency 0

The cross-over method

One of the easiest ways of working out formulas is to use the *cross-over method*. To use this method remember the following steps:

- Write down the symbols of the elements.
- Write the valencies below the symbols.
- Cancel down valencies if possible.
- Swap over the numbers and tidy up to give the formula.

Note that the number '1' is not written in formulas.

Example 1

Phosphorus chloride

Symbols	P	Cl
Valencies	3	1

Formula PCl_3

Example 2

Aluminium sulphide

Symbols	Al	S
Valencies	3	2

Formula Al_2S_3

Example 3

Carbon oxide

Symbols	C	O_2
Valencies	~~4~~ 2	~~2~~ 1

cancelled down

Formula CO_2

Example 4

Magnesium oxide

Symbols	Mg	O
Valencies	~~2~~ 1	~~2~~ 1

cancelled down

Formula MgO

Some compounds contain elements that have more than one valency. The names of these compounds contain roman numerals to indicate the valency the element.

Example 5

Iron(II) chloride

	Fe	Cl
Symbols	Fe	Cl
Valencies	2	1

Formula $FeCl_2$

Iron(III) chloride

	Fe	Cl
Symbols	Fe	Cl
Valencies	3	1

Formula $FeCl_3$

The most common roman numerals used in formulas are:

I = 1; II = 2; III = 3; IV = 4; V = 5; VI = 6 and VII = 7.

Odd formulas

The formulas of some compounds do not fit the valencies of their elements. In these compounds the name sometimes tell you the formula by adding a prefix in front of the element's name.

For example:

For example:	carbon monoxide	CO
	sulphur trioxide	SO_3
	dinitrogen tetraoxide	N_2O_4

Remember

The prefixes used in these 'odd' formulas are:

mono-	for **1**
di-	for **2**
tri-	for **3**
tetra-	for **4**
penta-	for **5**

State symbols

Sometimes you want to give information about the state of a substance, along with its formula.

Key Point

The symbols that are used to show the state of a substance are:

(s)	solid
(l)	liquid
(g)	gas
(aq)	dissolved in water

For example:	molten sulphur	S(l)
	sodium chloride solution	NaCl(aq)
	oxygen gas	O_2(g)
	ice (solid water)	H_2O(s)

Summary

- A molecule is a group of atoms held together by covalent bonds.
- Molecules and covalent bonds are usually formed by non-metals
- Sharing pairs of electrons between atoms forms the covalent bond.
- By sharing electrons the atoms become stable with electron arrangements like a noble gas.
- The covalent bond is an attraction between the protons and the shared electrons.
- A chemical formula tells you the ratio of elements in a compound.
- A molecular formula tells you the atoms in a molecule.
- The formula of most elements is just their symbol, however some contain diatomic molecules and are written H_2, I_2, O_2, etc.
- The formulas of main group compounds can be worked out from their valencies.
- Mono-, di-, tri-, tetra-, and penta- tell us the number of atoms in 'odd' formula.
- The state symbols used are (s), (l), (g) and (aq).

FUELS

Without the energy supplied by fossil fuels your life would be very different. But where do fossil fuels come form, what are they made of and what happens when fuels burn? This chapter looks at the chemistry of fuels. In particular it deals with the formation, refining and uses of crude oil.

Key Words

★ **acid rain** ★ **balanced equations** ★ **burning** ★ **catalytic converter** ★ **combustion** ★ **exothermic** ★ **finite** ★ **flammability** ★ **fossil fuel** ★ **fraction** ★ **fractional distillation** ★ **fuel** ★ **hydrocarbon** ★ **incomplete combustion** ★ **natural gas** ★ **pollution** ★ **symbol equation** ★ **viscosity** ★ **word equation**

Fuels and burning

In chemistry a **fuel** is a source of energy, which burns. Burning, or **combustion**, is a rapid reaction involving oxygen that gives out heat and light energy. Chemical reactions that give out heat are called **exothermic** reactions.

Remember

Air normally provides the oxygen for combustion.

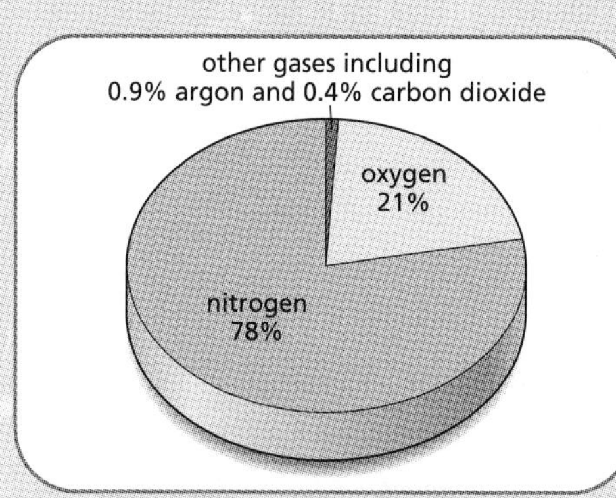

Figure 5.1 The composition of air

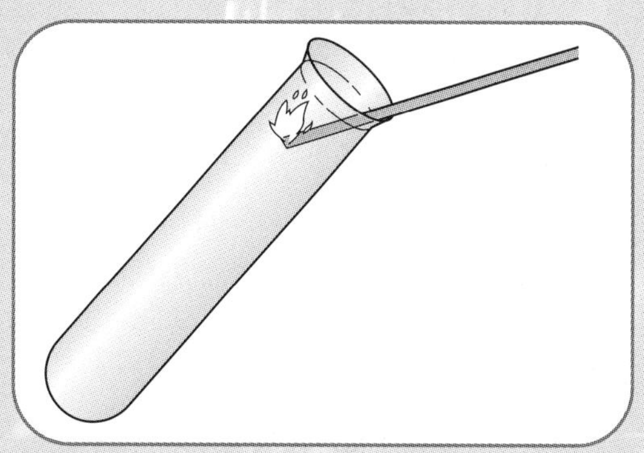

Figure 5.2 The test for oxygen is that it relights a glowing splint

Fossil fuels

Most of the fuels we use today come from coal, oil or natural gas. These are called **fossil fuels** as they have been formed from once living material.

Coal is formed from the remains of plants that died millions of years ago. Layers of earth and rock covered the vegetation, producing heat and pressure that changed the plant material into coal. Coal contains a mixture of chemicals, but is mainly made up of the element carbon.

Crude oil and **natural gas** are formed from the remains of tiny sea creatures. Like coal, they are formed underground by the action of heat and pressure over millions of years. Both crude oil and natural gas are made up of mixtures of chemicals called **hydrocarbons**.

Coal, oil and gas are all **finite** resources so they will eventually run out. This could result in a *fuel crisis*. To avoid running out of energy resources we will need to develop the use of renewable energies like wind, solar power and wave power. This could save our fossil fuel resources for other uses.

Hydrocarbons

Both crude oil and natural gas are mixtures of chemicals called **hydrocarbons**. These are compounds of carbon and hydrogen only. Crude oil is a complex mixture containing thousands of different hydrocarbon molecules.

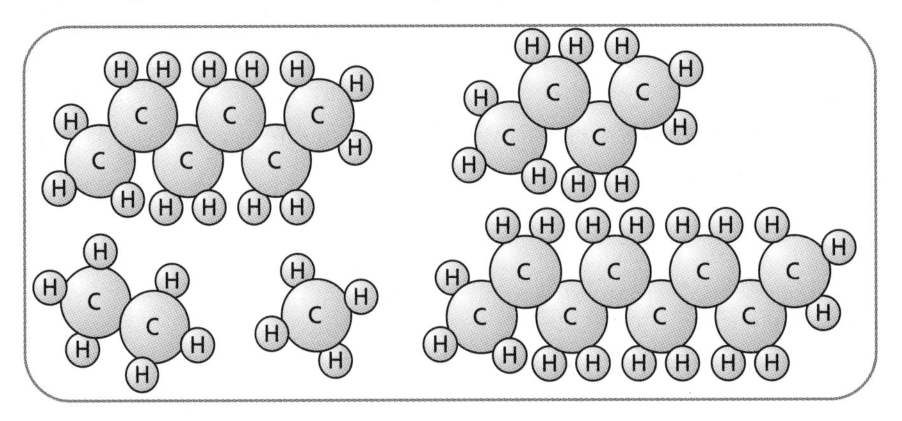

Figure 5.3 The hydrocarbons in oil and natural gas

Natural gas, which is usually found along with oil, is the main fuel gas for homes and industry. Natural gas is mostly methane (over 95%), mixed with small amounts of other gases. Methane is the simplest hydrocarbon, with one carbon atom and four hydrogen atoms in each molecule.

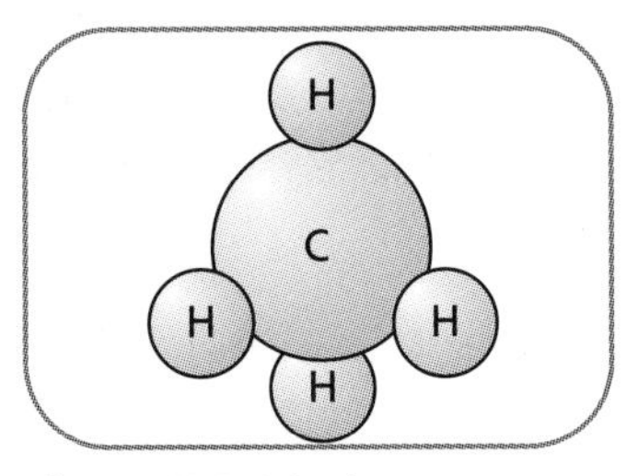

Figure 5.4 Methane, molecular formula CH_4

Products of combustion

The products of burning hydrocarbons can be found by experiment.

This experiment tells you that carbon dioxide and water are formed when hydrocarbons burn completely in oxygen or air.

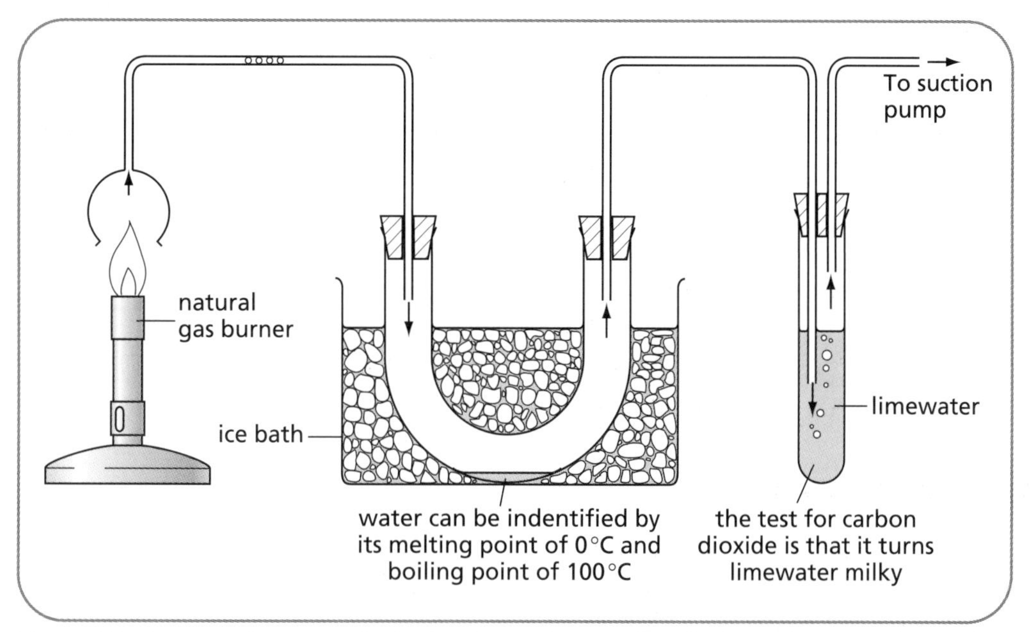

Figure 5.5 An experiment to find what is produced when hydrocarbons are burned

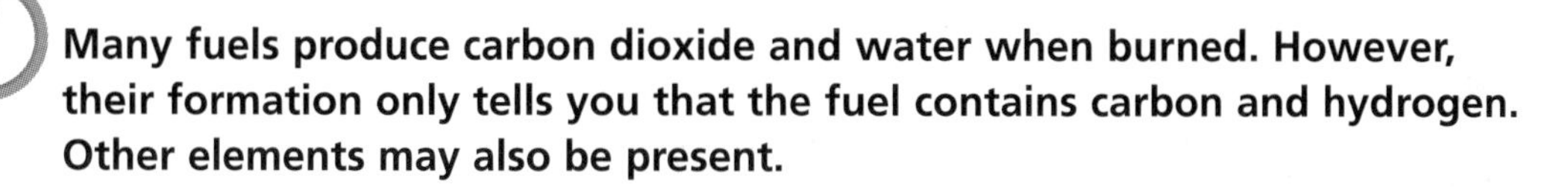

Many fuels produce carbon dioxide and water when burned. However, their formation only tells you that the fuel contains carbon and hydrogen. Other elements may also be present.

For example, when methane burns in oxygen, carbon dioxide and water are formed. This chemical change can be written as a **word equation**. In equations the chemicals at the start, the reactants, are written on the left, while the new substances formed, the products, are written on the right. The arrow in between shows the direction of the change.

word equation methane + oxygen → carbon dioxide + water

We often write equations using formulas in place of words. These **symbol equations** give more information about the chemical reaction.

symbol equation $CH_4 + O_2 \rightarrow CO_2 + H_2O$

Butane gas is another hydrocarbon (formula C_4H_{10}). It is used as lighter fuel. The word and symbol equations for the combustion of butane are as shown below

word equation butane + oxygen → carbon dioxide + water

symbol equation $C_4H_{10} + O_2 \rightarrow CO_2 + H_2O$

Equation To Learn

The general equation for the combustion of a hydrocarbon is:

hydrocarbon + **oxygen** → **carbon dioxide** + **water**

Oil refining

The crude oil, which is extracted from the ground, is a complex mixture of thousands of different hydrocarbons. These hydrocarbons vary in size from one to over a hundred carbon atoms per molecule. In this state crude oil is not very useful. However, once it is refined, many valuable products can be obtained. The first and most important process in oil refining is **fractional distillation**. This separates the crude oil into fractions by differences in boiling points.

Fractional distillation in industry

In industry fractional distillation is carried out in a **fractionating tower**. The crude oil is first heated until most of it has evaporated and turned into a gas. Then the mixture is piped into the bottom of the fractionating tower and as the gases rise up the tower they cool and start to condense back to liquids. This means that different hydrocarbon fractions can be taken off at different levels as the temperature decreases up the tower.

All substances have specific melting points and boiling points, which often depend on their molecular size.

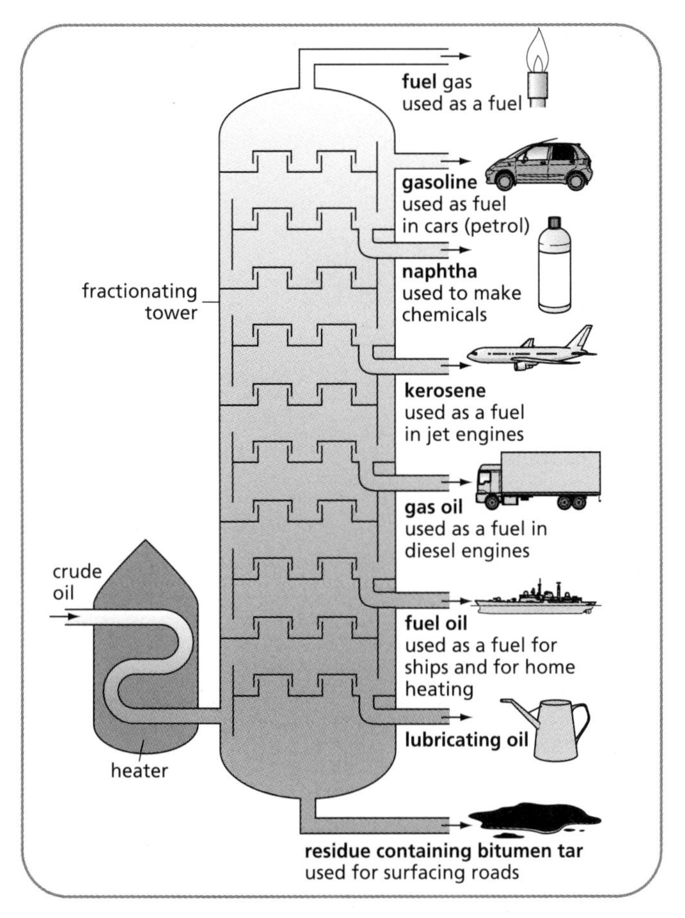

Figure 5.6 The fractionating tower

The residue will undergo further fractional distillation to produce heavier fractions like lubricating oils, waxes, fuel oils and bitumen tar.

Fractional distillation in the laboratory

Distillation can be carried out in the laboratory as shown below.

If the oil is heated slowly, the first hydrocarbons to evaporate and be collected are the fuel gases. Then other fractions are produced as the temperature rises. At the end of the distillation the highest boiling point fractions, like bitumen, are left in the test tube.

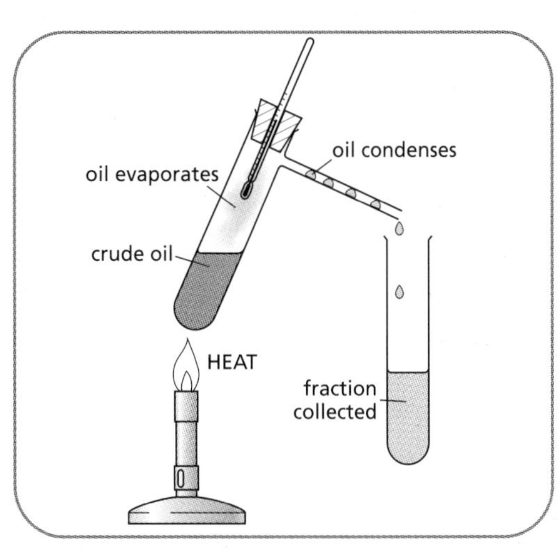

Figure 5.7 Oil distillation in the laboratory

About oil fractions

The different oil **fractions** are similar in that they all contain hydrocarbon molecules. However, they are different as the molecules are different sizes and they have different properties and uses.

Crude oil fraction	Number of carbon atoms	Uses of fractions
fuel gases	C_1 to C_4	Portable stoves and heaters
gasoline	C_4 to C_{10}	Petrol for cars
naphtha	C_6 to C_{12}	To make chemicals
kerosene	C_9 to C_{16}	Paraffin and fuel for jet aircraft
gas oil	C_{15} to C_{25}	Diesel fuel for lorries and trains
fuel oil	C_{20} to C_{36}	Domestic and industrial heating
lubricating oils	C_{34} to C_{70}	For all machinery
bitumen tar	$> C_{70}$	Tar for roads and roofs

As the *boiling point* of fractions increases:

- The *size of molecules* increase.
- The *ease of evaporation* decreases.
- The *flammability* increases.
- The *viscosity* increases.

Viscosity describes the thickness of a liquid, and how easily it runs.

Flammability describes how easily a substance catches fire.

Explaining properties

The lowest boiling point fractions are sometimes called the *lightest fractions*. They evaporate easily, have a low viscosity and are highly flammable. This is because they contain the smallest molecules, which are easier to separate and ignite.

The highest boiling point fractions, the *heaviest fractions*, do not evaporate easily, have a high viscosity and a low flammability. This is explained by the fact that they contain larger molecules, which are harder to separate and do not ignite easily.

Fuels and pollution

The extraction and use of fossil fuels often results in the **pollution** of our environment. The extraction of coal and oil are dangerous and dirty processes and oil spills at sea are a hazard to all sea life. However it is the combustion of coal and oil products in power stations and cars that causes the greatest pollution problems.

When fossil fuels burn they often produce sulphur dioxide (SO_2), as many fuels contain small amounts of sulphur. Sulphur dioxide is a problem as it is one of the main gases responsible for **acid rain**.

Motorcars pollute the atmosphere in several ways. Exhaust fumes can contain a mixture of pollutants such as:

- soot (C), carbon monoxide (CO) and unburned hydrocarbons (C_xH_y) due to incomplete combustion of the petrol and diesel fuels.
- Nitrogen dioxide (NO_2), another acid gas, is produced in petrol engines by the high-energy spark.
- Lead (Pb) compounds, which are added to some petrol, can also cause problems as they are poisonous.

Incomplete combustion is a major source of pollution. It occurs when any fuel burns in an insufficient supply of oxygen. This can happen in gas fires, water heaters or car engines. One of the products of incomplete combustion is carbon monoxide (CO). This is a poisonous gas, and is responsible for many deaths each year.

Figure 5.8 Pollution caused by cars

Reducing pollution

Reducing pollution from fuels is important to us all, and it can be achieved in several different ways.

- Removing sulphur compounds in fossil fuels before they are burnt reduces the SO_2 produced.
- Using car engines that are more efficient reduces incomplete combustion. This can be done by decreasing the fuel-to-air ratio. So there is less fuel and more oxygen to allow complete combustion.
- Using **catalytic converters** in car exhaust systems. These change harmful gases, like CO and NO_2, into harmless gases like CO_2 and N_2. The catalysts used in catalytic converters are expensive transition metals like platinum. The catalyst is spread over a honeycomb structure to give it a large surface area so it can convert more polluting gases.

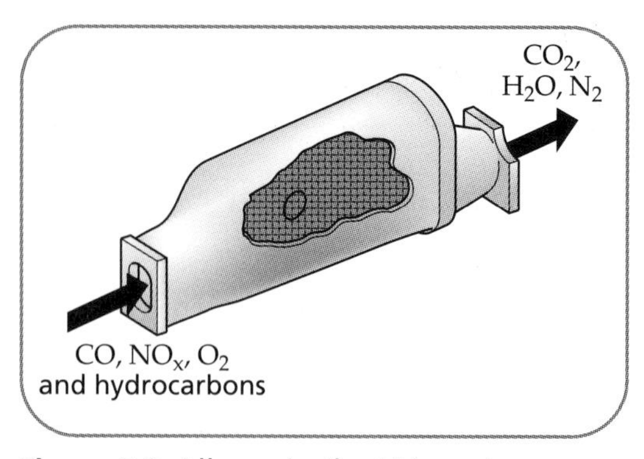

Figure 5.9 All cars in the UK must now have a catalytic converter to reduce pollution

Balanced chemical equations

A **word** or **symbol equation** describes what happens in a chemical reaction. Consider the equations for the combustion of hydrogen.

word equation	hydrogen + oxygen → water
symbol equation	$H_2 + O_2 \rightarrow H_2O$

In a chemical reaction, the reactants are what you start with and the products are the new substances formed.

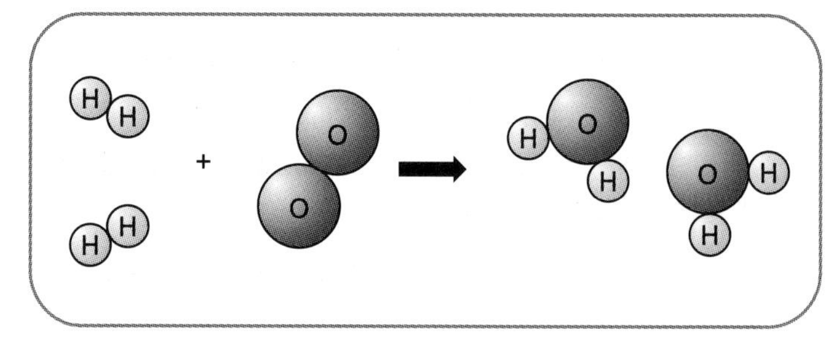

Figure 5.10 Molecules of water being formed

All chemical reactions involve rearranging atoms and molecules to form new substances. So atoms cannot be created or lost. This means that there should be the same number of atoms at the end as there was at the beginning. In the combustion of hydrogen two molecules of hydrogen are required for every oxygen molecule, and this forms two molecules of water.

To show that two molecules of a substance are needed, you put a 2 in *front* of the formula. This is how you balance the number of atoms on each side of the equation to make a **balanced equation**.

balanced equation $2H_2 + O_2 \rightarrow 2H_2O$

This equation is balanced, as there are the same numbers of atoms of hydrogen and oxygen on both sides.

Key Points

Remember when balancing equations:

- Count the atoms on each side.
- Put numbers in front of formulas to balance atoms.
- Do not change the formulas of substances.

Consider the combustion of methane.

symbol equation $CH_4 + O_2 \rightarrow CO_2 + H_2O$

To balance this equation, we need two water molecules on the right to balance the hydrogen atoms, with four on each side. Then you need two oxygen molecules on the left, so there will be four oxygen atoms on each side.

balanced equation $CH_4 + 2O_2 \rightarrow CO_2 + 2H_2O$

State symbols can be used in symbol or balanced equations to give more information about reactants and products.

For example, when solid carbon burns in a sufficient supply of oxygen gas, carbon dioxide gas is produced.

$$C(s) + O_2(g) \rightarrow CO_2(g)$$

However, the incomplete combustion of carbon produces carbon monoxide gas.

$$2C(s) + O_2(g) \rightarrow 2CO(g)$$

Summary

- A fuel is a substance that burns giving out energy.
- Combustion (burning) is a reaction with oxygen, which gives out energy.
- Air is made up of 78% nitrogen and 20% oxygen.
- The test for oxygen is that it relights a glowing splint.
- Coal, oil and natural gas are fossil fuels formed from ancient living things.
- All fossil fuels are formed from the remains of living things which have been buried underground over millions of years.
- The fossil fuels are finite, this means they will eventually run out.
- Oil is a mixture of hydrocarbons, compounds containing hydrogen and carbon only.
- Distillation separates crude oil into fractions depending on their boiling points.
- The fractions in order of increasing boiling point are: fuel gases < gasoline < naphtha < kerosene < gas oil < fuel oil < lubricating oils < bitumen tar.
- As the boiling points of the fractions increase their molecular size increases.
- As molecular size increases their viscosity (the 'thickness' of liquid) increases.
- As molecular size increases their flammability (ease of catching fire) decreases.
- When hydrocarbons burn completely in oxygen they form carbon dioxide and water.
- The test for carbon dioxide is it turns limewater milky.
- Carbon monoxide and carbon (soot) are formed by incomplete combustion, because of a lack of oxygen.
- The pollution formed by burning fuels includes sulphur dioxide, nitrogen dioxide, soot and unburned hydrocarbons.
- Catalytic converters in car exhausts change harmful gases into harmless gases.
- Reducing the fuel-to-air ratio in car engines means more complete combustion.
- The balanced equation for a chemical reaction has the same number of atoms of each element on each side.

Chapter 6

HYDROCARBONS

Crude oil and the hydrocarbons it contains are amongst our most precious resources. But what are the properties of these hydrocarbons, and how can we make sense of them when there are thousands of different examples? This chapter deals with the structure and properties of the most common hydrocarbons. It also introduces catalytic cracking, an important process in oil refining.

Key Words

★ **addition reaction** ★ **alkane** ★ **alkene** ★ **catalytic cracking** ★ **cycloalkane** ★ **double bond** ★ **general formula** ★ **homologous series** ★ **isomer** ★ **saturated hydrocarbon** ★ **structural formula** ★ **unsaturated hydrocarbon**

Hydrocarbon series

The simplest hydrocarbon is called methane. It can be obtained from crude oil and makes up more than 95% of natural gas.

The other hydrocarbons found in crude oil can be made by joining different numbers of carbon atoms to form a chain of atoms. The next simplest hydrocarbon, which is called ethane, has two carbon atoms in each molecule.

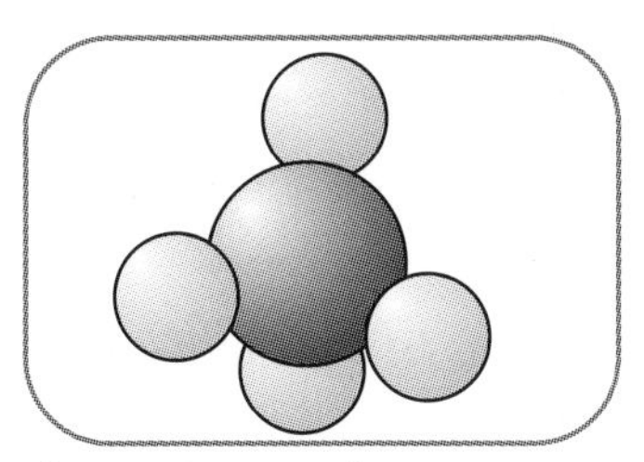

Figure 6.1 Methane, CH_4

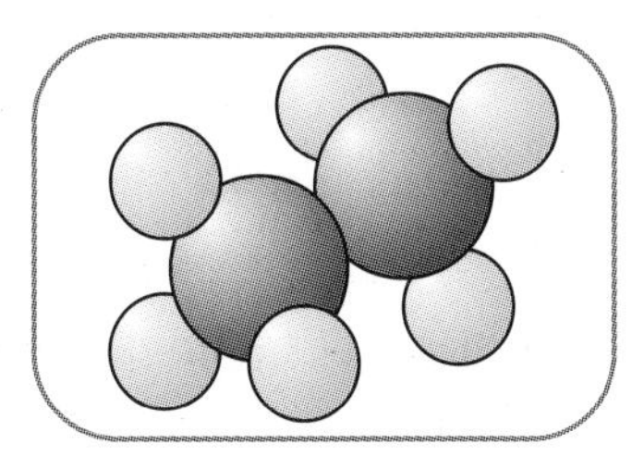

Figure 6.2 Ethane, C_2H_6

The hydrocarbon with three carbon atoms is called propane.

By adding more carbon atoms to the chain a series of hydrocarbons is formed. This series is called the **alkanes**. The following table shows the names, formulas and **structural formulas** of the first eight alkanes.

Names and formulas of first eight alkanes

Name	Molecular formula	Full structural formula	Physical state at 25°C
methane	CH_4	H–C(H)(H)–H	gas
ethane	C_2H_6	H–C(H)(H)–C(H)(H)–H	gas
propane	C_3H_8	H–C(H)(H)–C(H)(H)–C(H)(H)–H	gas
butane	C_4H_{10}	H–C(H)(H)–C(H)(H)–C(H)(H)–C(H)(H)–H	gas
pentane	C_5H_{12}	H–C(H)(H)–C(H)(H)–C(H)(H)–C(H)(H)–C(H)(H)–H	liquid
hexane	C_6H_{14}	H–C(H)(H)–C(H)(H)–C(H)(H)–C(H)(H)–C(H)(H)–C(H)(H)–H	liquid
heptane	C_7H_{16}	H–C(H)(H)–C(H)(H)–C(H)(H)–C(H)(H)–C(H)(H)–C(H)(H)–C(H)(H)–H	liquid
octane	C_8H_{18}	H–C(H)(H)–C(H)(H)–C(H)(H)–C(H)(H)–C(H)(H)–C(H)(H)–C(H)(H)–C(H)(H)–H	liquid

Notice that the formulas of alkanes fit a pattern and there is a **general formula** for the series.

Key Point

General formula of **alkanes** is C_nH_{2n+2}

The full structural formula of a hydrocarbon does not show the actual shape of the molecules. Instead it is a 'flattened' out, for easy drawing. The full structural formula does, however, show which atoms are joined together.

Structures, properties and uses of alkanes

The alkanes are described as **saturated hydrocarbons** as they are made up of chains of carbon atoms held together by single C–C bonds. Each carbon atom forms four bonds while each hydrogen atom forms only one bond The shapes of alkane molecules are based on the tetrahedral structure found in methane.

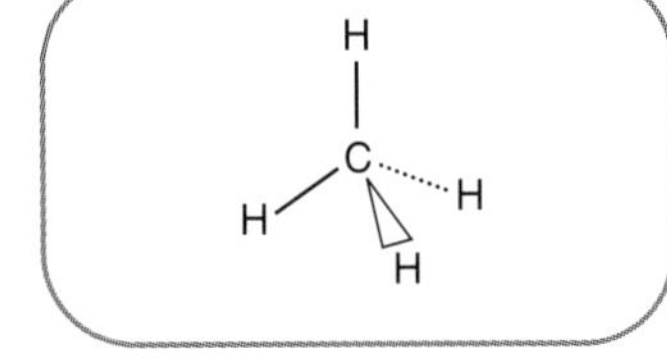

Figure 6.3 All alkanes have a tetrahedral structure like methane

This shape is caused by repulsions between the negative electrons in the bonds. Thus a tetrahedral structure is as far apart as the bonds can get.

The lower alkanes have similar chemical and physical properties. In general they:

- Are colourless liquids and gases.
- Have low boiling points.
- Are very flammable.
- Form carbon dioxide and water on complete combustion.

Alkanes are found in many fuels made from crude oil. For example, fuel gases, petrol, diesel and paraffin all contain high proportions of alkanes. The equations for burning alkanes are similar, as they all produce carbon dioxide and water on complete combustion.

Equation To Learn

alkane + oxygen → carbon dioxide + water

The equations below represents the combustion of hexane.

balanced equation	$C_6H_{14} + 9\frac{1}{2}O_2 \rightarrow 6CO_2 + 7H_2O$
word equation	hexane + oxygen → carbon dioxide + water

As you go down the alkane series their boiling points increase, as the molecular size increases. From methane to butane (C_1 to C_4) they are gases at room temperature. Larger alkanes are liquids.

Homologous series

The alkanes are an example of a **homologous series**. That is a series of compounds that

- Fit a general formula.
- Have similar chemical properties
- Show a gradual change in physical properties, like boiling points, viscosity, etc.

Homologous series are important in chemistry as they help us understand the properties and reactions of large groups of compounds.

The boiling points increase regularly as you go down the group of alkanes

Alkane	Boiling point (°C)
methane	–164
ethane	–89
propane	–42
butane	–1
pentane	36
hexane	69
heptane	98
octane	126

Cycloalkanes

The **cycloalkanes** are another homologous series of hydrocarbons. A cycloalkane is similar to an alkane except its molecules contain a ring of the carbon atoms. The simplest cycloalkane is cyclopropane.

Names and formulas of first four cycloalkanes

Name	Molecular formula	Full structural formula	Shortened structural formula
cyclopropane	C_3H_6	H H C H–C–C–H H H	CH_2 CH_2—CH_2
cyclobutane	C_4H_8	H H H–C–C–H H–C–C–H H H	CH_2—CH_2 CH_2—CH_2

Name	Molecular formula	Full structural formula	Shortened structural formula
cyclopentane	C_5H_{10}		CH_2 CH_2 CH_2 CH_2—CH_2
cyclohexane	C_6H_{12}		CH_2 CH_2 CH_2 CH_2 CH_2 CH_2

Note that the formulas of the cycloalkanes also fit a pattern and there is a general formula for the series.

Key Point

General formula of **cycloalkanes** is C_nH_{2n}

The cycloalkanes are similar to alkanes in properties. They are generally low boiling point, colourless liquids and gases. They are **saturated hydrocarbons** as they only contain C–C **single bonds** and they are flammable, producing carbon dioxide and water on complete combustion.

balanced equation	C_4H_8	+	$6O_2$	$\rightarrow$	$4CO_2$	+	$4H_2O$
word equation	cyclobutane	+	oxygen	$\rightarrow$	carbon dioxide	+	water

Alkenes

There is a third homologous series of hydrocarbons called **alkenes**. The alkenes are similar to alkanes but contain a **double bond** between a pair of carbon atoms. The simplest alkene is ethene.

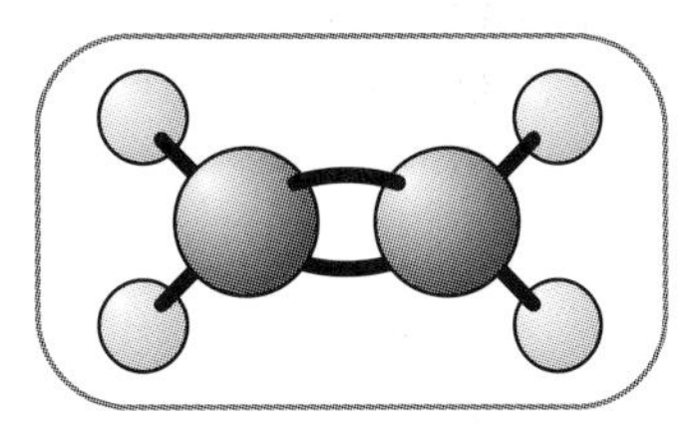

Figure 6.4 Ethene, C_2H_4

In the series of alkenes the carbons are joined in a chain with one C=C double bond in each molecule.

Names and formulas of first four alkenes

Name	Molecular formula	Full structural formula	Shortened structural formula
ethene	C_2H_4	H₂C=CH₂ (all bonds shown)	$CH_2{=}CH_2$
propene	C_3H_6	H–C(H)(H)–C(H)=C(H)(H)	$CH_3CH{=}CH_2$
butene	C_4H_8	H–C(H)(H)–C(H)(H)–C(H)=C(H)(H)	$CH_3CH_2CH{=}CH_2$
pentene	C_5H_{10}	H–C(H)(H)–C(H)(H)–C(H)(H)–C(H)=C(H)(H)	$CH_3CH_2CH_2CH{=}CH_2$

Note that the shortened structural formulas do not show all the bonds, just the groups of atoms that are joined together. If asked for any structural formula, give the full structural formula, as they are easier to draw.

The general formula for the alkenes and cycloalkanes is the same. Each has two fewer hydrogen atoms than the corresponding alkane.

Key Point

General formula of **alkenes** is C_nH_{2n}

In many ways alkenes are similar to alkanes and cycloalkanes in properties. They are all low boiling point, colourless liquids and gases. They are also flammable, producing carbon dioxide and water on complete combustion.

balanced equation $C_2H_4 + 3O_2 \rightarrow 2CO_2 + 2H_2O$

word equation ethene + oxygen → carbon dioxide + water

They are however described as **unsaturated hydrocarbons** as they contain a C=C **double bond**. This gives them some different properties from the saturated hydrocarbons.

Saturated or unsaturated?

All hydrocarbon series, alkane, cycloalkane and alkenes, have many properties in common: low boiling point, colourless, flammable, etc. As all hydrocarbons are similar it is difficult to distinguish one group from another.

The one hydrocarbon reaction that is different is the reaction with bromine water. The diagram below shows what happens when bromine water is added to an alkane, cycloalkane and alkene.

Bromine water is a dilute solution of the element bromine. It is light brown in colour, and is written as $Br_2(aq)$.

Alkanes and cycloalkanes, the saturated hydrocarbons, do not react quickly with bromine water, which remains brown in colour.

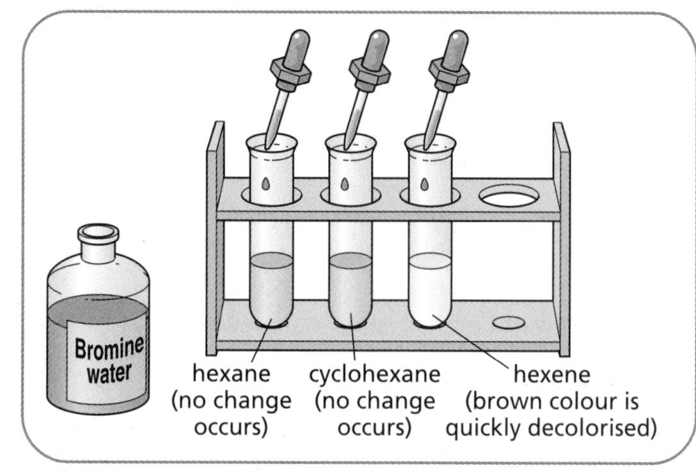

Figure 6.5 Adding bromine water to an alkane, a cycloalkane and an alkene

Alkenes, the unsaturated hydrocarbons, react quickly with bromine water decolorising it, so the brown colour disappears.

Note: The test for unsaturation (alkenes) is they quickly decolorise bromine water.

Isomers

Some hydrocarbon molecules, with the same numbers of number of carbon and hydrogen atoms, have their atoms joined together in different ways. These are called **isomers**.

Isomers are compounds with the same molecular formulas but different structural formulas.

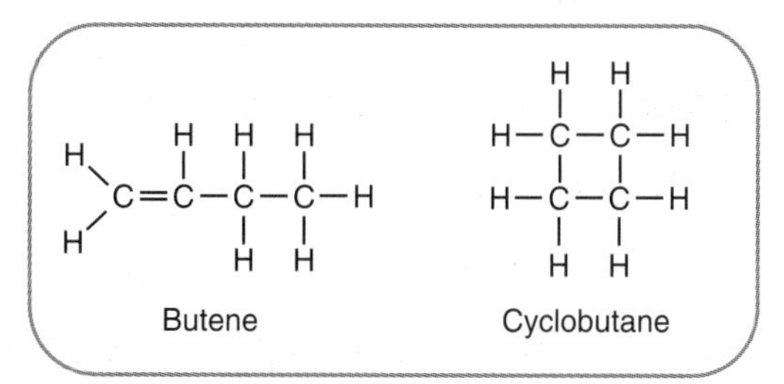

Figure 6.6 Butene and cyclobutane are isomers; both have the formula C_4H_8

```
                                  H   H   H
                                  |   |   |
     H   H   H   H            H - C - C - C - H
     |   |   |   |                |   |   |
 H - C - C - C - C - H            H   |   H
     |   |   |   |                    |
     H   H   H   H                H - C - H
                                      |
                                      H
```

Figure 6.7 Butane has two isomers

The larger the molecule the greater the number of possible isomers. For example, C_5H_{12} has three isomer while $C_{25}H_{52}$ has 25 352 788 isomers!

Addition reactions

Unsaturated hydrocarbons like alkenes take part in **addition reactions**. In these reactions the double bond in the alkene partially breaks and an atom is added on to each of the carbons which formed the C=C double bond.

The reaction between alkenes and bromine water is an example of an addition reaction.

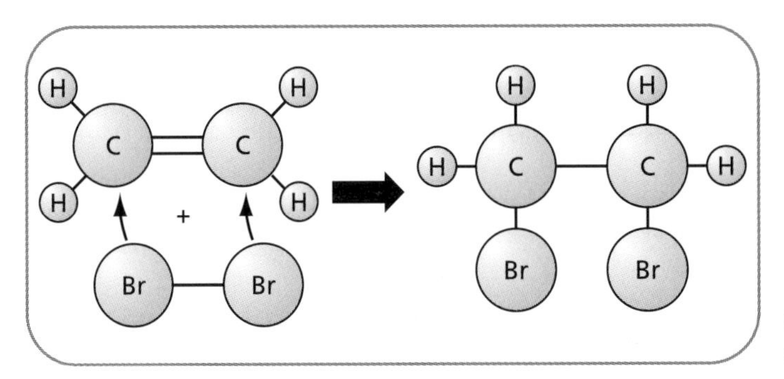

Figure 6.8 The addition of bromine to ethene

The equation for the addition reaction between ethene and bromine is shown below.

$$C_2H_4 + Br_2 \rightarrow C_2H_4Br_2$$

brown colourless

The test for unsaturation or alkenes is an example of an addition reaction. When alkenes quickly decolorise bromine water the bromine adds across the double bond.

Hydrogen can also add on to an alkene, across the double bond, to make an alkane.

The equation for this addition reaction is:

$$C_2H_4 + H_2 \rightarrow C_2H_6$$

ethene + hydrogen → ethane

Notice that in all addition reactions the double bond is removed and the product formed is saturated. Also, when hydrogen is added to an alkene, the corresponding alkane is formed.

Catalytic cracking

The crude oil from different countries contains different proportions of the oil fractions. This can be a problem, as there is an almost fixed demand for each of the fractions obtained. Generally we use more of the lighter fractions, like petrol. Therefore most crude oils contain more of the heavier fractions than are needed.

To meet the demand for the lighter fractions, we have to break up the larger molecule fractions that are not needed. The process is called **cracking** and it can be brought about by heating the heavy hydrocarbon fractions. This shakes up the large molecules, causing them to break apart into smaller groups of atoms.

In industry, and the laboratory, a **catalyst** of **aluminium oxide** can be used to speed up the reaction. Using a catalyst allows the reaction to take place at a lower temperature. In the oil refinery **catalytic cracking** is much more economic than cracking by heat alone.

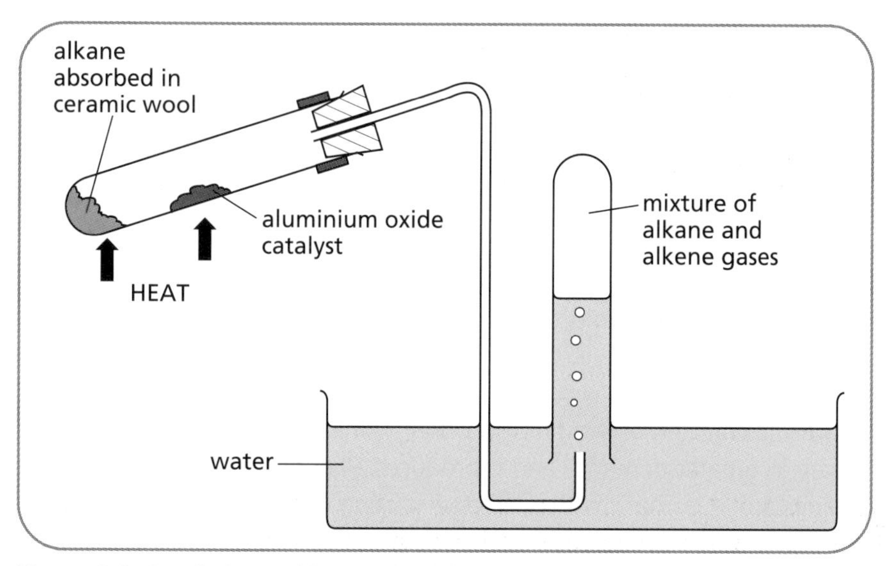

Figure 6.9 Catalytic cracking in the laboratory

When a liquid alkane is 'cracked' a mixture of gases is formed. Testing the properties of these gases tells you something about their structure. For example, the 'cracked' gases are flammable and quickly decolorise bromine water. This tells you that the products must contain smaller hydrocarbon molecules and at least one must be unsaturated.

An example of what happens during the cracking of an alkane is shown below.

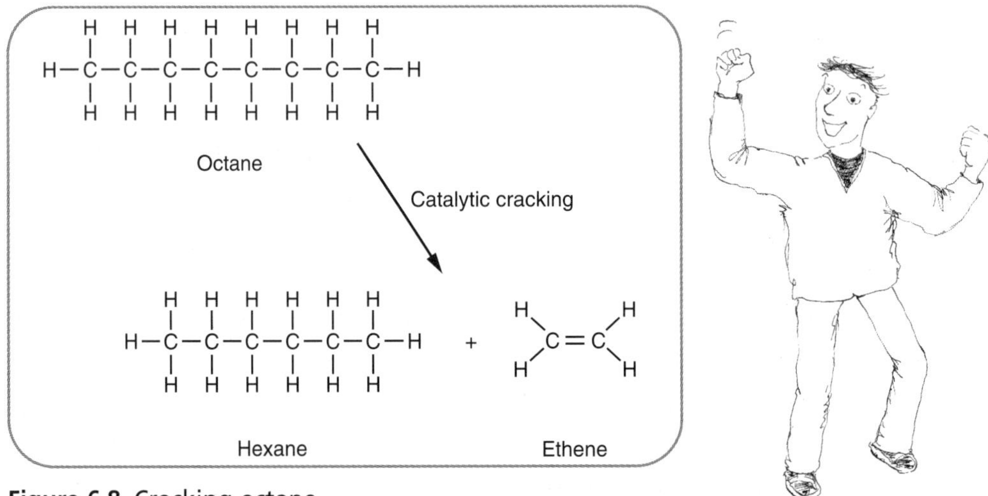

Figure 6.8 Cracking octane

In an equation for cracking, the number of carbon and hydrogen atoms must balance.

Example

$$C_{12}H_{26} \rightarrow C_6H_{14} + C_4H_8 + C_2H_4$$

This equation is balanced, as there are 12 carbon atoms and 26 hydrogen atoms on each side.

Note that although cracking can form a mixture of different products, it always produces at least one **alkene** molecule.

Importance of cracking

In industry catalytic cracking serves two purposes. It makes the lighter hydrocarbon fuels, which are in greater demand, and it produces alkenes for use in the chemical industry. Alkenes are important as they are the starting materials in the manufacture of many other chemicals, including **plastics**. (See Chapter 14 for more details about how plastics are made.)

Summary

- Hydrocarbons are molecular compounds of carbon and hydrogen only.
- Most hydrocarbons are colourless liquids and gases, which are flammable.
- Natural gas is mainly methane, CH_4, the simplest hydrocarbon.
- During complete combustion: hydrocarbon + oxygen → carbon dioxide + water.
- The start of the name tells you the number of carbon atoms. From C_1 to C_8, the names start: meth . . , eth . . , prop . . , but . . , pent . . , hex . . , hept . . and oct . . .
- The compounds in a homologous series fit a general formula and have similar chemical reactions.
- Alkanes: name ending . . ane; C_nH_{2n+2}; are saturated with only C–C single bonds.
- Cyclolkanes: name cyclo . . . . ane; C_nH_{2n}; are saturated with only C–C single bonds joined in a ring.
- Alkenes: name ending . . ene; C_nH_{2n}; are unsaturated with one C=C double bond.
- Test for unsaturated hydrocarbons (alkenes) is they quickly decolorise bromine water.
- Alkenes take part in addition reactions with Br_2, H_2, HBr, etc. The products are saturated.
- Isomers have the same molecular formulas but different structures.
- Catalytic cracking (catalyst aluminium oxide) breaks down long-chain hydrocarbons.
- Cracking produces smaller hydrocarbon molecules, for fractions in greater demand, and alkenes, for the plastics industry.

Chapter 7

PROPERTIES AND BONDING

Covalent bonds are formed when non-metal atoms share electrons. But are all bonds formed in the same way, and how can we explain the properties of substances with our theories on bonding? In this chapter modern ideas about bonding are developed by investigating properties like electrical conductivity, melting point, boiling point and solubility.

Key Words

★ bonds ★ conductivity ★ covalent bond ★ electrolysis ★ electrolyte ★ ionic bond ★ ion–electron equation ★ lattice structure ★ metallic bond ★ molecular structure ★ network structure ★ noble gas ★ stable electron arrangement

Conductors and insulators

Substances which allow electricity to pass through them are called **conductors**. Substances which do not allow electricity to pass are called non-conductors or **insulators**. The property of electrical **conductivity** is very useful when investigating the bonding and structure of substances.

When the conductivities of a range of different substances are tested, they appear to fall into three main groups.

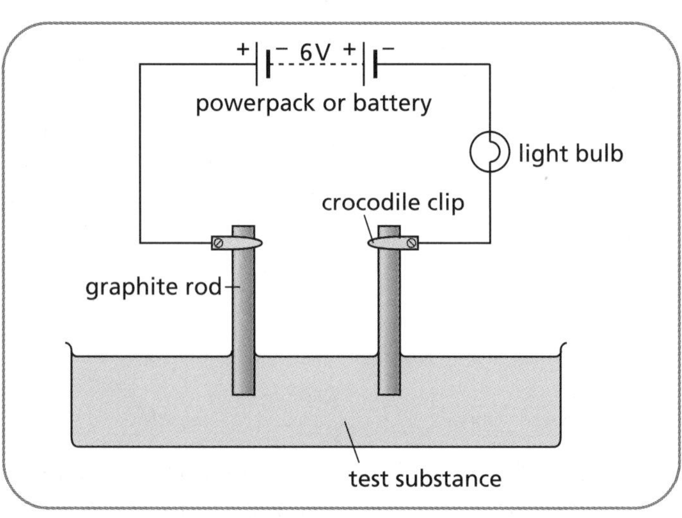

Figure 7.1 Testing electrical conductivity

Substances that are conductors when liquid and solid	Substances that are non-conductors when liquid or solid	Substances that are conductors when liquid or dissolved in water but are non-conductors when solid
copper	sulphur	sodium chloride
iron	oxygen	copper chloride
mercury	bromine	lead bromide
magnesium	carbon dioxide	potassium iodide
nickel	pentane	aluminium oxide
carbon (graphite)	silicon dioxide	nickel sulphate
This group contains metals and carbon in the form of *graphite*	This group contains non-metal elements and compounds	This group contains compounds formed between metals and non-metals

An electric current is a flow of moving charged particles. Therefore to conduct electricity a substance must contain freely moving charged particles. Bonding theories should explain the different conductivity properties of substances

Metallic bonding

Apart from graphite, **metals** are the only common solids that conduct electricity. It is thought that metals conduct because the electrons in their outer shells are freely moving. The passage of electricity does not change the metal: no chemical reaction occurs during conduction.

As the metal atoms lose their outer electrons this produces a regular structure of positive ions in a sea of free electrons. This is called a **metallic lattice structure**. The **metallic bond** that holds the lattice together is a result of attractions between the positive ions and the freely moving electrons.

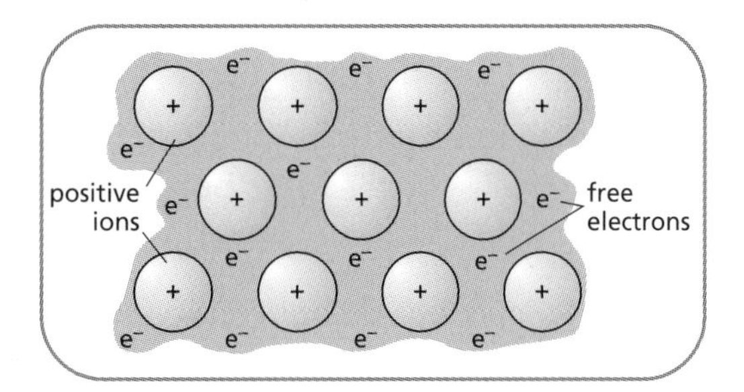

Figure 7.2 The metallic lattice structure

Ionic bonding

Compounds formed between metals and non-metals contain ions and are called **ionic compounds**. Ions are atoms or groups of atoms that have become charged due to the loss or gain of electrons.

Metals, which usually have one, two or three electrons in their outer shell, tend to **lose electrons** and form **positive ions**.

Non-metals, which usually have four, five, six or seven electrons in the outer shell, tend to **gain electrons** and form **negative ions**.

For example, when sodium chloride is formed the sodium atom loses an electron to the chlorine atom.

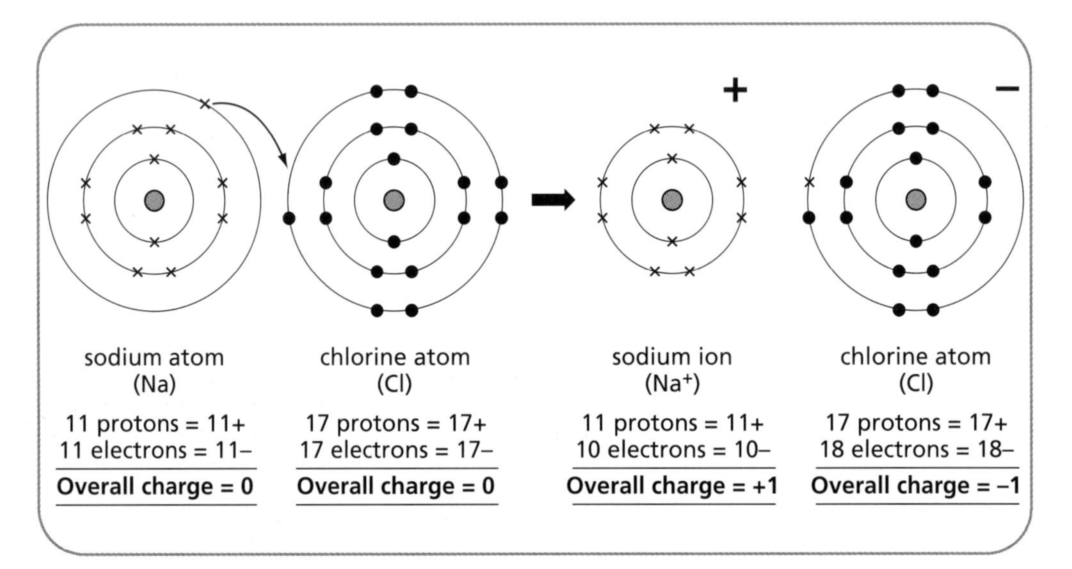

Figure 7.3 How an ionic bond is formed

By losing one electron the sodium atom forms a positive sodium ion and by gaining one electron the chlorine atom forms a negative chloride ion. Once made, the oppositely charged ions attract each other, and produce a giant regular structure called an **ionic lattice**. The strong attractions between billions of oppositely charged ions are the **ionic bonds** that hold the ionic lattice together.

Ionic compounds do not conduct electricity when solid, as the ions cannot move through the lattice structure. However, they can conduct when molten or in solution as the lattice is broken up and the ions can move. This process is called **electrolysis** and it causes the ionic compound to be broken up into its elements.

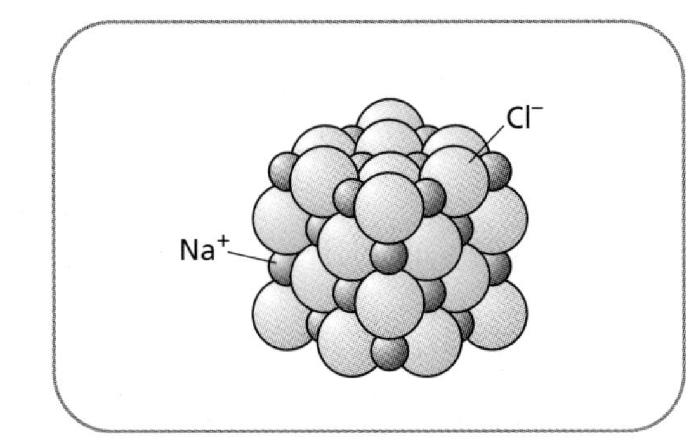

Figure 7.4 The sodium chloride lattice structure

During **electrolysis** the ions move towards the oppositely charged electrode. The positive ions, usually metal ions, move towards the negative electrode where they change into metal atoms. The negative ions, usually non-metal ions, move towards the positive electrode where they change into non-metal atoms. Thus the ionic compound is broken down to its elements by electrolysis.

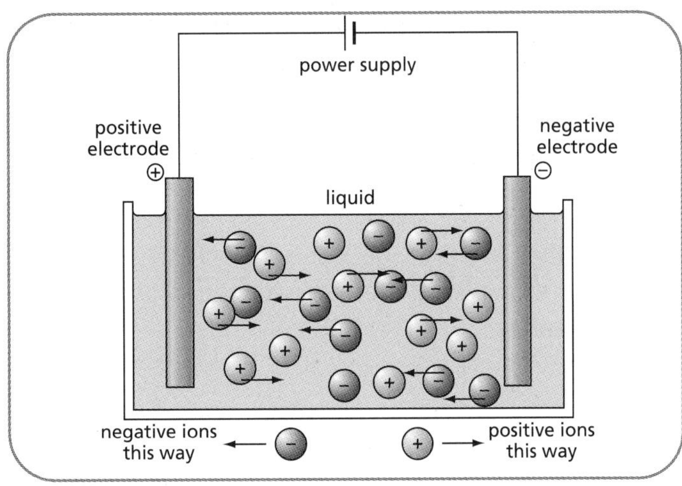

Figure 7.4 Electrolysis

Examples of electrolysis products

Molten compound being electrolysed	Product at the negative electrode	Product at the positive electrode
copper chloride	copper	chlorine
lead bromide	lead	bromine
aluminium oxide	aluminium	oxygen

A d.c. supply must be used during electrolysis to form specific products at each electrode. This is because a direct current keeps the ions moving in one direction.

More about ions

The type of ion formed by an element depends on the electron arrangement of its atoms. In general the electrons are lost or gained so that the ion has a **stable electron arrangement** like a **noble gas**. Some examples of the electron arrangements of atoms and ions are shown below.

Element	Atom symbol	Atom electron arrangement	Electrons lost or gained	Ion symbol	Ion electron arrangement
magnesium	Mg	2,8,2	loses 2e	Mg^{2+}	2,8
sulphur	S	2,8,6	gains 2e	S^{2-}	2,8,8
aluminium	Al	2,8,3	loses 3e	Al^{3+}	2.8
bromine	Br	2,8,18,7	gains 1e	Br^{-}	2,8,18,8

The loss and gain of electrons can be represented by an **ion–electron equation**. For example, the formation of the aluminium and bromide ions, described in the table above, could be written:

$$Al(s) \rightarrow Al^{3+}(aq) + 3e^-$$

$$Br_2 \rightarrow 2Br^- + 2e^-$$

Notice that two bromide ions are involved, as bromine is a diatomic element. (Further examples of these equations can be found on page 7 of the *Chemistry Data Booklet*.)

Covalent bonding

Non-metal elements and compounds usually have a molecular structure held together by **covalent bonds**. The bond is formed by sharing pairs of electrons between the atoms. By forming covalent bonds the atoms get a stable electron arrangement like a noble gas. The atoms in a covalent bond are held together by the attraction of the positive nuclei to the shared negative electrons. (See Chapter 4 for more about covalent bonding.)

Covalent elements and compounds are generally non-conductors of electricity. They do not contain free electrons or ions, so there are no freely moving charged particles that can carry the current in covalent substances.

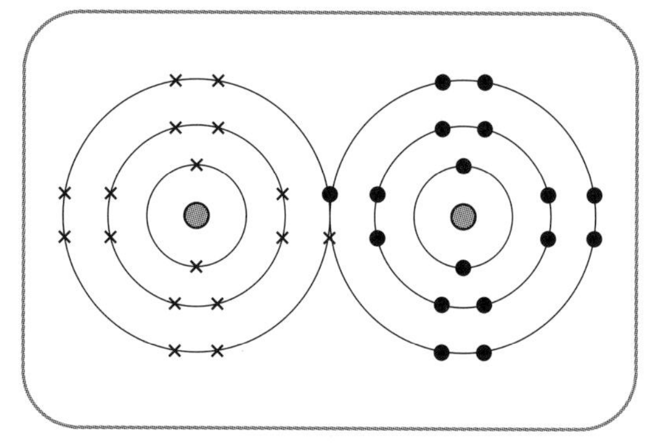

Figure 7.5 Chlorine, a typical covalent molecular structure

The only non-metal element to conduct electricity is carbon in the form of graphite. The other main form of carbon, diamond, is however a non-conductor.

Ionic formula

In ionic compounds metals tend to form positive ions and non-metals tend to form negative ions. The number of charges on the ion is the same as the valency of the element. These charges are sometimes shown in the formulas of ionic compounds.

Example 1

Aluminium sulphide

Symbols and charges Al^{3+} S^{2-}

Valencies 3 2

Ionic formula $(Al^{3+})_2\ (S^{2-})_3$

Example 2

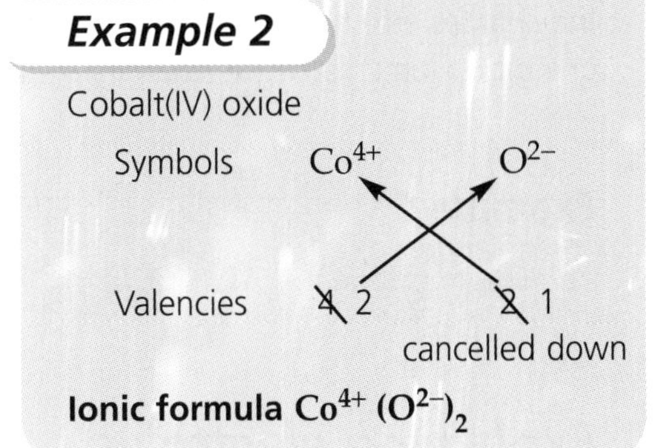

Cobalt(IV) oxide

Symbols Co^{4+} O^{2-}

Valencies ~~4~~ 2 ~~2~~ 1

cancelled down

Ionic formula $Co^{4+}\ (O^{2-})_2$

Some ions contain more than one atom. For example, sulphate is SO_4^{2-} and hydroxide is OH^-.

The formulas for these 'group' ions can be found in the data booklet. The formulas for compounds containing these ions are worked out in the same way as before. The valency of ions which contain more than one atom is still the charge number.

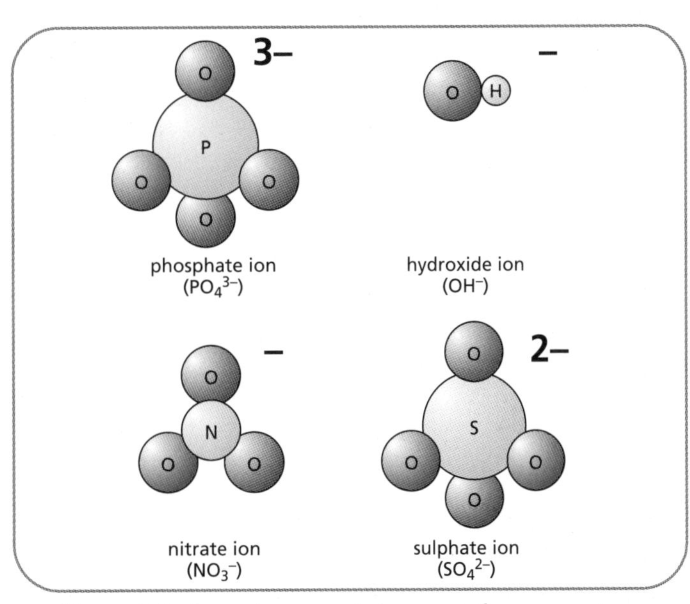

Figure 7.6 Some ions contain more than one atom

Example 3

Ammonium sulphate

Symbols and charges NH_4^+ SO_4^{2-}

Valencies 1 2

Ionic formula $(NH_4^+)_2\ SO_4^{2-}$

Example 4

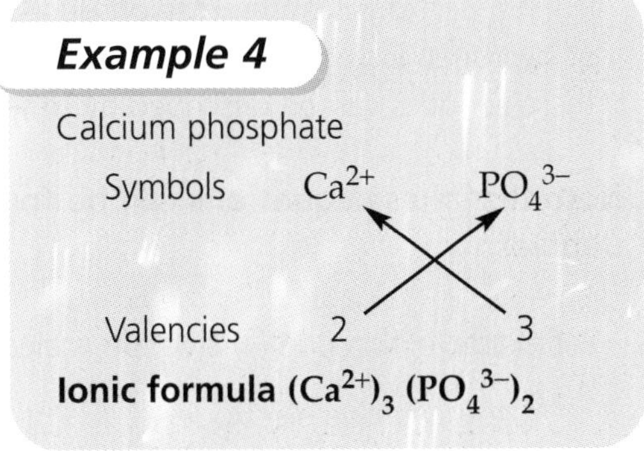

Calcium phosphate

Symbols Ca^{2+} PO_4^{3-}

Valencies 2 3

Ionic formula $(Ca^{2+})_3\ (PO_4^{3-})_2$

Notice that brackets ( ) are used in formulas that contain more than one ion, when the charges are shown. Brackets are also used if the formula contains more than one group ion, even if the charges are not shown.

Example 5

Aluminium carbonate: $Al_2(CO_3)_3$

Example 6

Magnesium nitrate: $Mg(NO_3)_2$

Explaining electrolysis

Electrolysis involves passing an electric current through an **electrolyte**. That is, a solution that contains ions or a molten ionic compound. Ionic substances can only conduct when molten or in solution, as in these states the lattice is broken up and the ions can move. During electrolysis the electrical energy breaks down the electrolyte and elements are formed at the electrodes. The diagram below shows what happens during the electrolysis of molten lead bromide.

The **negative electrode** attracts the lead ions.

The lead ions gain electrons at the electrode:

$$Pb^{2+}(aq) + 2e^- \rightarrow Pb(s)$$

So silvery beads of lead metal are formed.

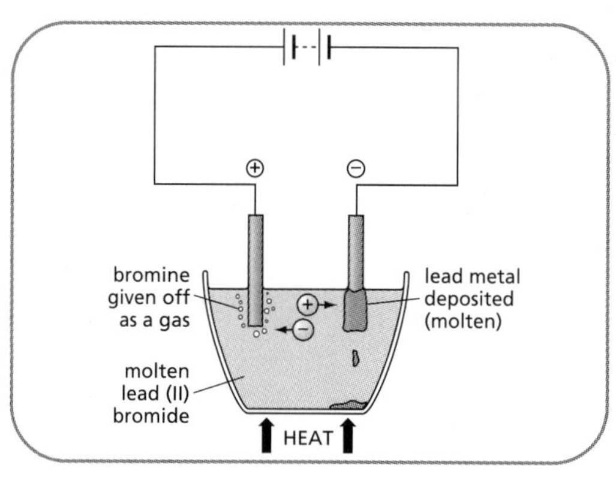

Figure 7.7 Electrolysis of molten lead bromide

The **positive electrode** attracts the bromide ions.

The bromide ions lose electrons at the electrode

$$2Br^-(aq) \rightarrow Br_2(aq) + 2e^-$$

So brown fumes of bromine gas are formed.

During electrolysis non-metal ions lose electrons at the positive electrode and metal or hydrogen ions gain electrons at the negative electrode. The changes occurring at the electrodes can be described by **ion–electron equations**.

Note that these equations can be found on page 7 of the *Chemistry Data Booklet*.

For example, if copper(II) chloride is electrolysed the reactions are:

At the negative electrode $Cu^{2+}(aq) + 2e^- \rightarrow Cu(s)$

At the positive electrode $2Cl^-(aq) \rightarrow Cl_2(g) + 2e^-$

Coloured ions

Some ionic compounds are coloured due to the presence of coloured ions.

- Copper(II) ions (Cu^{2+}) are blue.
- Dichromate ions ($Cr_2O_7^{2-}$) are orange.

Coloured ions can be used to show the movement of ions during electrolysis.

Explaining melting points and boiling points

The state of a substance at room temperature depends on its melting point and boiling point, which in turn depend on its bonding and structure.

All **ionic compounds** have high melting points and boiling points and are always solids at room temperature.

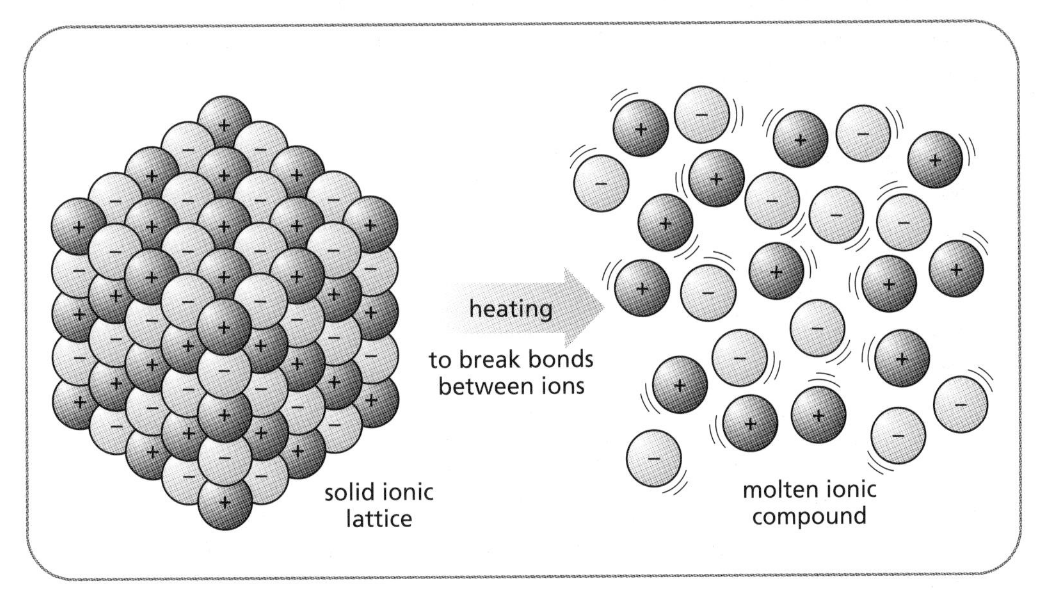

Figure 7.8 Melting an ionic compound

This is because the ionic lattice is held together by billions of strong ionic bonds, and so it takes a lot of energy to separate the ions and change the solid into a liquid

The **metallic lattice** is also held together by billions of strong metallic bonds. So the melting points of metals are also generally high, and most metals are solids at room temperature. (The exception is mercury, which is a liquid.)

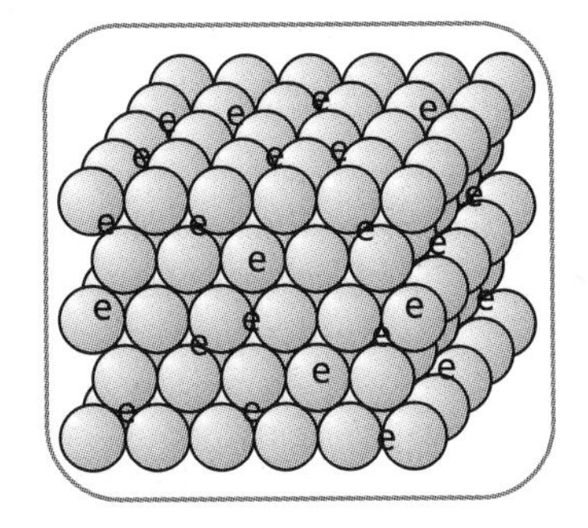

Figure 7.9 Metallic lattice

Most **covalent elements** and **compounds** have low melting points and boiling points as they have a **molecular structure**. The atoms are held together in the molecules by strong covalent bonds, but the forces of attraction between the molecules are weak.

When a molecular substance melts no covalent bonds are broken, the molecules are just separated from each other. This doesn't usually take a lot of energy, so most covalent substances are low melting point solids, liquids or gases.

Figure 7.10 A covalent molecular substance

There are, however, a few covalent elements and compounds that have very high melting points. These substances are not made up of molecules but have a covalent lattice or **network structure**. The most common examples are carbon as diamond (melting point 3827 °C) and silicon dioxide (melting point 1610 °C) which is the main chemical in sand and glass.

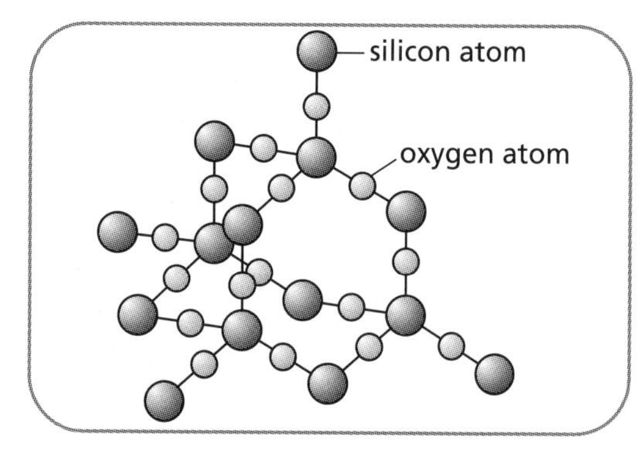

Figure 7.11 Silicon dioxide covalent lattice structure

These have very high melting points and boiling points as billions of strong covalent bonds have to be broken when they change state.

Solubility and bonding

A **solution** is formed when a **solute** dissolve in a **solvent**. In a solution the solute particles are mixed in between the solvent particles so you cannot tell them apart.

Ionic compounds are often soluble in water. When an ionic compound dissolves, the lattice is broken up and all the ions spread out and mix in between the water molecules.

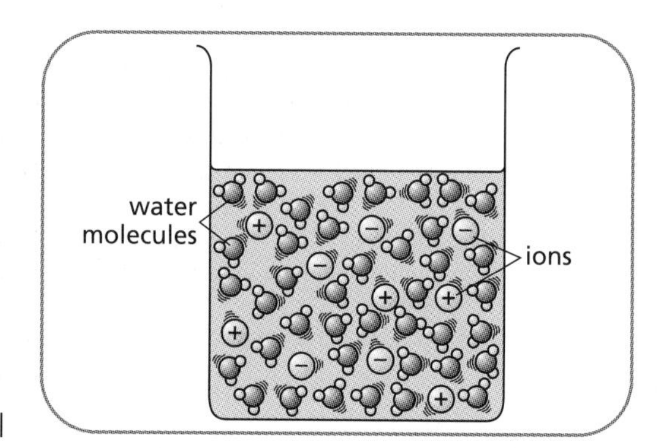

Figure 7.12 Ionic compound dissolving in water

Covalent substances are usually not soluble in water but can dissolve in other solvents. For example, wax is insoluble in water but dissolves in alkane mixtures like white spirit.

Comparing bonding types

To understand the properties of matter you need to understand how the atoms are held together. The following table compares the most important aspects of ionic, covalent and metallic bonding.

Type of bonding	Ionic bonding	Covalent bonding	Metallic bonding
Where formed	Compounds of metals and non-metals	Non-metal elements and compounds	Metals and alloys
How formed	Metals lose electrons and non-metals gain electrons to form ions	Pairs of outer electrons are shared between atoms	Outer electrons of metal atoms are free to move between atoms
Structure	All have an ionic lattice structure Opposite charged ions attract each other	Most have a covalent molecular structure Some have a covalent network	All have a metallic lattice structure e e e e e e e e e e Held together by attraction between positive ions and free electrons
General properties	◆ Conduct electricity when liquid and in solution as ions can move ◆ Non-conductors when solid as ions cannot move ◆ High m.p. and b.p., all solids, as lattice structure ◆ Generally soluble in water	◆ Non-conductors of electricity in any state, as no free ions or electrons ◆ Covalent molecules have low m.p. and b.p. as little attraction between molecules ◆ Covalent networks have high m.p. and b.p. as lattice structure ◆ Generally insoluble in water, dissolve in other solvents	◆ Conduct electricity when solid and liquid as outer electrons free to move through lattice ◆ Fairly high m.p. and b.p., mostly solids, as lattice structure ◆ Insoluble in any solvents
Examples	sodium chloride iron(III) oxide calcium sulphate zinc nitrate copper(II) sulphate	carbon dioxide methane iodine (all molecular) carbon silicon dioxide (both lattices)	copper aluminium gold steel Brass

Summary

- Bonds are the forces of attraction, which hold atoms together in molecules or lattices.
- In metals, metallic bonding involves free outer electrons and a lattice structure.
- Metals conduct when solid and liquid because of the free outer electrons.
- In compounds of metals and non-metals, ionic bonding involves the transfer of electrons and lattice structures.
- Ionic compounds are non-conductors when solid, but conduct when liquid or in solution, when the ions are free to move.
- In non-metals, covalent bonding involves sharing electrons, and usually molecular structures. A few (e.g. C and SiO_2) have a lattice structure.
- Covalent substances are non-conductors in any state, as there are no free charged particles. (Exception: carbon as graphite)
- In general, covalent molecular structures have low melting points, while all lattice structures have high melting points.
- Electrolysis involves passing an electric current through an electrolyte.
- An electrolyte contains ions that can move, e.g. a solution or molten ionic compound.
- During electrolysis the electrolyte is broken down to its elements.
- As metal ions are positive a metal is usually formed at the negative electrode.
- As non-metal ions are negative a non-metal is formed at the positive electrode.
- Ion–electron equations describe what happens during electrolysis (examples in data booklet).
- Some ions are coloured (e.g. Cu^{2+} is blue, $Cr_2O_7^{2-}$ is orange). These can be seen moving during electrolysis.
- Ionic compounds tend to dissolve in water while covalent substances will dissolve in other solvents.

Chapter 8

ACIDS AND ALKALIS

Most people think that acids and alkalis are dangerous liquids. However, is that all you need to know about them? This topic looks at the properties of acids and alkalis in the laboratory, in industry and in our environment. Some chemical calculations, involving concentrations of solutions, are also introduced during this chapter.

Key Words

★ **acid** ★ **acidic oxide** ★ **acid rain** ★ **alkali** ★ **concentration**
★ **formula mass** ★ **gram formula mass** ★ **indicator** ★ **mole** ★ **neutral**
★ **pH scale** ★ **universal indicator**

Acids, alkalis and the pH scale

All aqueous solutions, that is solutions in water, are either acidic, alkaline or neutral. You can tell them apart by using **indicators** which change colour depending on the type of solution present. Some examples of common **acids**, **alkalis** and **neutral** solutions, which can be found in and around the home, are shown below.

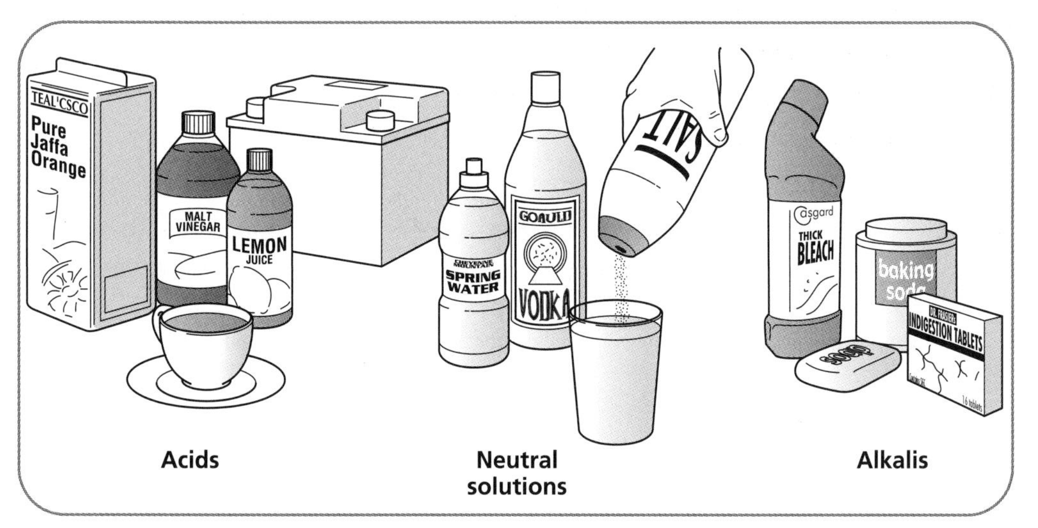

Figure 8.1 Some common acids, alkalis and neutral solutions found in the home

The **pH scale** measures how acidic or alkaline a solution is. The scale runs from below 0 to 14 and above. Neutral solutions have a pH of 7. Acids have a pH less than 7 and alkalis have a pH greater than 7. The pH scale and colours of **universal indicator** are shown below.

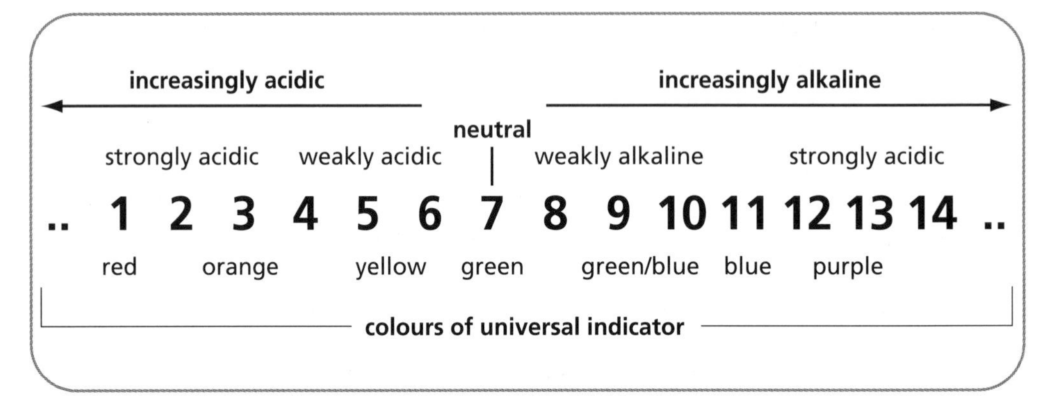

Figure 8.2 The pH scale

Making acids and alkalis

Acids and alkalis can be formed from the oxides of elements if they are soluble in water.

Soluble **metal oxides** dissolve to form **alkalis**.

For example, Li_2O, Na_2O and CaO all form alkaline solutions.

Soluble **non-metal oxides** dissolve to form **acids**.

For example, SO_2, CO_2 and NO_2 all form acidic solutions.

Note that the solubility of many metal oxides can be found in the data booklet.

Acid rain

Rainwater is naturally acidic, with a pH of about 5.5 This is mainly due to dissolved non-metal oxides, like carbon dioxide. The burning of fossil fuels, in power stations and cars, produces sulphur dioxide and nitrogen dioxide. As these are soluble non-metal oxides, they dissolve in the rainwater to make it more acidic than normal with a pH of less than 5. This is called '**acid rain**' and it damages buildings, plants and life in rivers and lochs. (See Chapter 9 for more about the effects of acid rain.)

Water and solutions

Pure water (H_2O) is a covalent molecular compound and should therefore not conduct electricity. However, experiments have shown that water does conduct slightly, so it must contain a few ions.

Further investigations have revealed that the ions present in water are hydrogen ions (H^+) and hydroxide ions (OH^-). The ions are formed by the break up of a few water molecules. This can be shown in an equation:

$$H_2O(l) \rightleftharpoons H^+(aq) + OH^-(aq)$$

This means that pure water, and all neutral solutions, will contain a small number of H^+ and OH^- ions in equal concentrations.

During the electrolysis of water the H^+ and OH^- ions carry the current as the water is broken up into its elements. Hydrogen is formed at the negative electrode and oxygen is formed at the positive electrode.

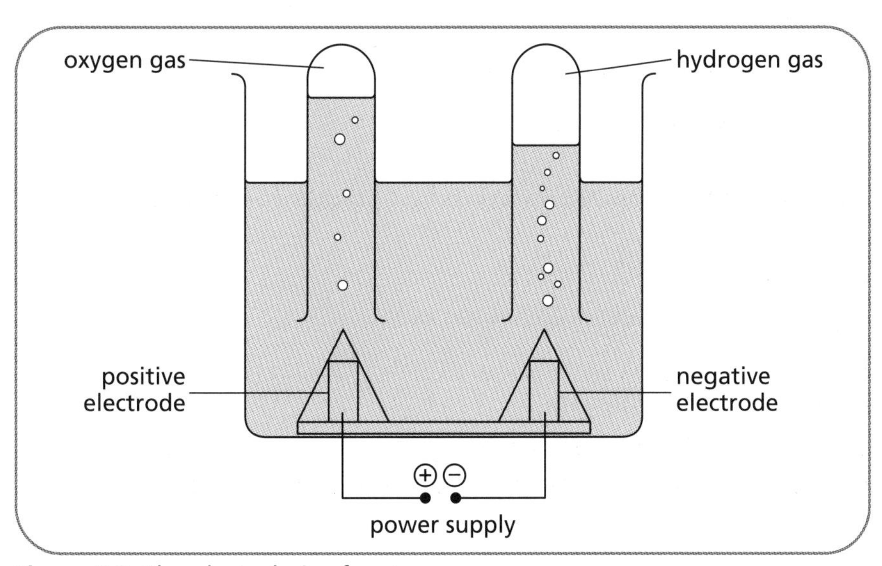

Figure 8.3 The electrolysis of water

In contrast most acids and alkalis are good conductors of electricity, which means they must contain fairly high concentrations of ions.

When an acid solution is electrolysed, hydrogen gas is always formed at the negative electrode. This is because all **acids** contain **excess H^+ ions**. Some examples of common laboratory acids are shown in the table.

Name of acid	Formula	Ionic formula
Hydrochloric acid	HCl	$H^+(aq) + Cl^-(aq)$
Sulphuric acid	H_2SO_4	$2H^+(aq) + SO_4^{2-}(aq)$
Nitric acid	HNO_3	$H^+(aq) + NO_3^-(aq)$
Phosphoric acid	H_3PO_4	$3H^+(aq) + PO_4^{3-}(aq)$

Further experiments showed that all **alkaline** solutions contained **excess OH^- ions**. Some examples of common laboratory alkalis are shown in the table.

Name of alkali	Formula	Ionic formula
Sodium hydroxide	$NaOH$	$Na^+(aq) + OH^-(aq)$
Potassium hydroxide	KOH	$K^+(aq) + OH^-(aq)$
Calcium hydroxide	$Ca(OH)_2$	$Ca^{2+}(aq) + 2OH^-(aq)$

The ionic formulas for solutions are sometimes written to show the ions separated, as they would be when dissolved in water.

Key Points

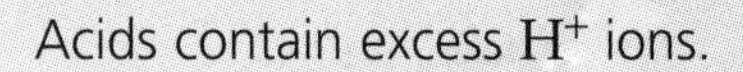

Acids contain excess H^+ ions.

Alkalis contain excess OH^- ions.

Neutral solutions contain the same number of H^+ and OH^- ions.

Diluting acids and alkalis

When a solution is diluted water is added to lower the concentration of solute.

When an acidic solution is diluted, the concentration of hydrogen ions (H^+) must decrease, so the acidity decreases and the pH rises towards 7.

When an alkaline solution is diluted, the concentration of hydroxide ions (OH^-) must decrease, so the alkalinity decreases and the pH falls towards 7.

Introduction to calculations, formula mass and the mole

The Credit examination will include some questions on chemical calculations. In these calculations we often use quantities called the **formula mass** and the **mole**.

The **formula mass** of a substance is the sum of the **relative atomic masses** of all the atoms in its formula. The units used for relative atomic masses and formula masses are 'atomic mass units' or 'amu' for short.

A **mole** of any substance is a quantity equal to its formula mass in grams. This is sometimes called the **gram formula mass**. We should be able to calculate the mass of a given number of moles and the number of moles in any given mass of a substance.

Example 1

What is the formula mass of sodium carbonate?

Formula Na_2CO_3

$3 \times 16 = 48$

$1 \times 2 = 12$

$2 \times 23 = 46$

Total $= 106$

$\therefore$ Formula mass = 106 amu

Example 2

What is the mass of 1 mole of sodium hydroxide?

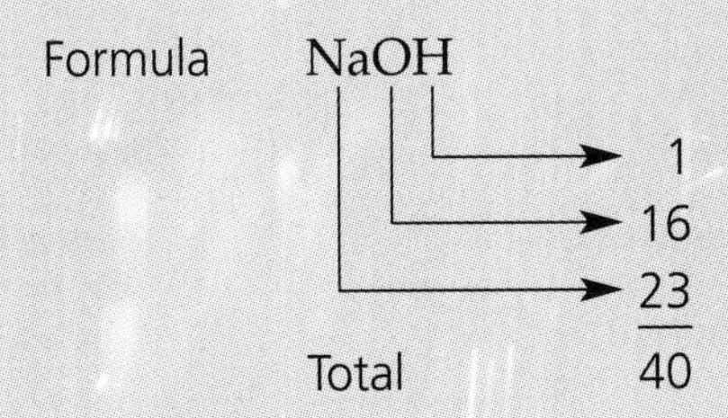

$\therefore$ Formula mass = 40 amu

$\therefore$ 1 mole = 40 g

Example 3

What is the mass of 2.5 moles of iron metal?

Formula of iron = Fe
Formula mass = 56 amu
∴ 1 mole = 56 g
∴ 2.5 moles = 2.5 × 56 = 140 g
∴ 2.5 moles Fe = 140 g

Example 4

How many moles of sulphuric acid are in 4.9 g of the acid?

Formula of sulphuric acid = H_2SO_4

4 × 16 = 64
1 × 32 = 32
2 × 1 = 2
Total = 98

∴ Formula mass = 98 amu

∴ 1 mole = 98 g

∴ 4.9g $= \frac{4.9}{98} = 0.05$ moles

∴ 4.9 g of sulphuric acid = 0.05 moles

Calculating concentrations

The **concentration** of a **solution** is a measure of the amount of solute dissolved in a certain volume of water. In chemistry we usually measure concentration in **moles per litre** or **mol/l** for short. Calculations involving concentrations can be carried out by using the following 'equation triangle'.

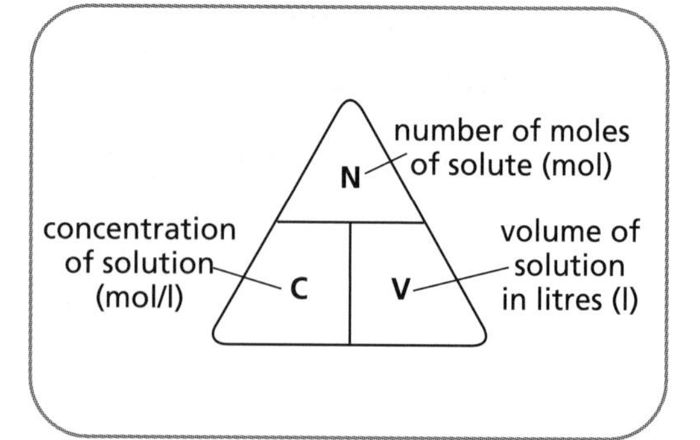

Figure 8.4 By covering up one letter you can see the relationship to the other two. For example, if we cover C you see that: $C = \frac{N}{V}$

Example 1

What mass of lithium hydroxide is dissolved in 200 cm^3 of a solution with a concentration of 2.0 mol/l? (The formula mass of lithium hydroxide is 24 amu.)

From the equation triangle N = C × V

$\therefore$ N = 2.0 × 0.2 (remember the volume is in litres)

$\therefore$ N = 0.4 moles

1 mole of LiOH = 24 g

$\therefore$ 0.4 moles = 0.4 × 24 = 9.6 g

$\therefore$ Mass of LiOH = 9.6 g

Remember

To change cm^3 into litres, divide by 1000.

Example 2

A pupil made up a solution of nitric acid (HNO_3) by dissolving 12.6 g of the acid in water and making the total volume up to 2500 cm^3. What is the concentration of this acid solution? (The formula mass of nitric acid is 63 amu.)

First calculate the number of moles of acid.

1 mole of nitric acid = 63 g

$\therefore 12.6\,g = \frac{12.6}{63} = 0.2$ moles

From the equation triangle $C = \frac{N}{V}$

$\therefore C = \frac{0.2}{2.5}$ (remember the volume is in litres)

$\therefore$ C = 0.08 mol/l

$\therefore$ Concentration of nitric acid = 0.08 mol/l

Summary

- Solutions can be acidic, alkaline or neutral; test with an indicator.
- Common acids include vinegar and fruit juice; common alkalis include bleach and baking soda.
- The pH scale: 0 → 7 is acidic; 7 is neutral; 7 → 14 is alkaline.
- Soluble non-metal oxides form acids, while soluble metal oxides form alkalis.
- Common acids: HCl, H_2SO_4, HNO_3 and H_3PO_4 all contain excess H^+ ions.
- All acids produce hydrogen gas at the negative electrode during electrolysis.
- The test for hydrogen is it burns with a pop.
- Common alkalis: $NaOH$, $LiOH$ and $Ca(OH)_2$ all contain excess OH^- ions.
- Pure water is neutral as it contains the same number of H^+ and OH^- ions.
- Acid rain is caused by SO_2 and NO_2 from the burning fossil fuels, dissolving in rain.
- If acids are diluted, acidity decreases as pH rises towards 7.
- If alkalis are diluted, alkalinity decreases as pH falls towards 7.
- The formula mass of a substance is the sum of all its relative atomic masses.
- A mole of a substance is its formula mass in grams.
- The concentration of a solution is measured in moles per litre (mol/l).
- The number of moles, concentration or volume of a solution can be calculated using $N = C \times V$.

Chapter 9

REACTIONS OF ACIDS

The chemical term 'neutralisation' is commonly used when referring to problems like acid indigestion, soil pH and acid rain. But do you really understand what is happening during neutralisation? For example, do you know what products are formed or what kind of chemical reaction occurs? This chapter looks at some of the reactions of acids, including different neutralisation reactions. It also deals with the formation of salts and calculations from titrations.

Key Words

★ **acid** ★ **alkali** ★ **base** ★ **neutralisation** ★ **neutraliser** ★ **precipitation** ★ **salt** ★ **spectator ion** ★ **titration**

Neutralisation

In simple terms, **neutralisation** is the reaction between an acid and a **neutraliser**. During neutralisation the acid is destroyed, and the pH of the solution gets closer to 7.

Everyday examples of neutralisation include:

- Adding lime (calcium oxide) to acid soils to help certain plants grow well.
- Taking 'milk of magnesia' (magnesium hydroxide) to treat acid indigestion.
- Using lime (calcium oxide) or limestone (calcium carbonate) to reduce acidity in lochs.

Bases and alkalis

Metal oxides, **metal hydroxides** and **metal carbonates** are all examples of neutralisers. In chemistry these substances are called **bases** and if the base is soluble in water it is called an **alkali**.

Examples of bases and alkalis

Examples of insoluble bases	Examples of soluble alkalis
copper(II) oxide	lithium oxide
lead(II) oxide	sodium hydroxide
iron(III) oxide	potassium hydroxide
aluminium hydroxide	sodium carbonate
nickel(II) hydroxide	
calcium carbonate	
zinc carbonate	

The reaction between an **acid** and a **base**, or an **alkali**, is a **neutralisation** reaction. During neutralisation a new substance called a **salt**, and **water** are always formed. As the acid reacts the solution becomes less acidic, so the pH rises towards 7.

Note that in chemistry the term 'salt' is used to describe a group of compounds formed from acids. It does not just refer to the common salt that we use to flavour our food.

Key Points

The type of salt formed during neutralisation depends on the acid used.

- Hydrochloric acid forms chloride salts.
- Sulphuric acid forms sulphate salts.
- Nitric acid forms nitrate salts.

Acid reactions in the laboratory

The neutralisation of acids using metal oxides, metal hydroxides and metal carbonates can be easily carried out in the laboratory.

Acids and metal oxides

The reaction between an acid and a metal oxide always produces a salt and water.

Equation To Learn

acid + metal oxide → salt + water

For example, if copper(II) oxide is heated with some sulphuric acid solution, the following reaction occurs.

$$H_2SO_4 \quad + \quad CuO \quad \rightarrow \quad CuSO_4 \quad + \quad H_2O$$

sulphuric acid + copper(II) oxide → copper(II) sulphate + water

The salt produced, copper(II) sulphate, is blue so it dissolves to form a blue solution, while the water just mixes with the rest of the solution.

Acids and metal hydroxides

The reaction between an acid and a metal hydroxide also always produces a salt and water.

Equation To Learn

acid + metal hydroxide → salt + water

For example, if sodium hydroxide solution is added to a solution of hydrochloric acid, the following reaction occurs.

$$NaOH \quad + \quad HCl \quad \rightarrow \quad NaCl \quad + \quad H_2O$$

sodium hydroxide + hydrochloric acid → sodium chloride + water

The salt produced, sodium chloride, is white so it dissolve to form a colourless solution. As there is no visible sign of a reaction, an indicator has to be used to show when the acid is exactly neutralised. As the alkali is added the pH rises towards 7. The same reaction occurs if the acid is added to the alkali; however, in this case the pH falls towards 7.

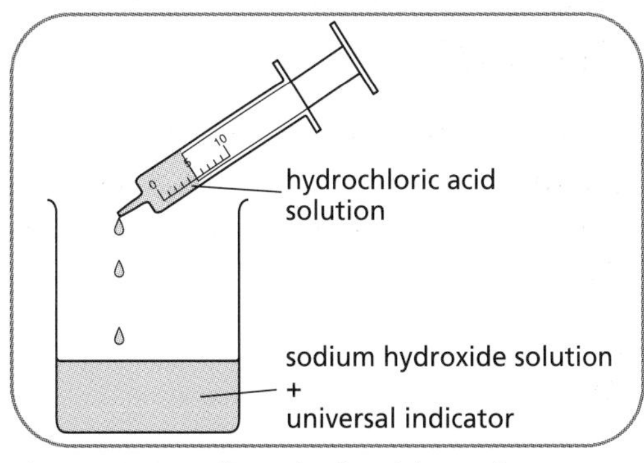

Figure 9.1 Sodium hydroxide and hydrochloric acid

Acids and metal carbonates

The reaction between an acid and a metal carbonate always produces a salt, water and carbon dioxide gas.

Equation To Learn

acid + metal carbonate → salt + water + carbon dioxide

For example, if zinc(II) carbonate is added to some nitric acid solution, the following reaction occurs.

$$2HNO_3 + ZnCO_3 \rightarrow Zn(NO_3)_2 + H_2O + CO_2$$

nitric acid + zinc(II) carbonate → zinc(II) nitrate + water + carbon dioxide

The salt produced, zinc(II) nitrate, is white so it dissolves to form a colourless solution. **Effervescence** (fizzing) occurs as carbon dioxide gas is also produced.

Making salts

Salts are formed during neutralisation when the hydrogen ion (H^+) of the acid is replaced with another positive ion. The actual salt formed depends on the base and the acid that is used. Some examples of the salts formed from hydrochloric acid are shown opposite.

There are two common ways of making a sample of a particular salt. The method used depends on the solubility of the salt.

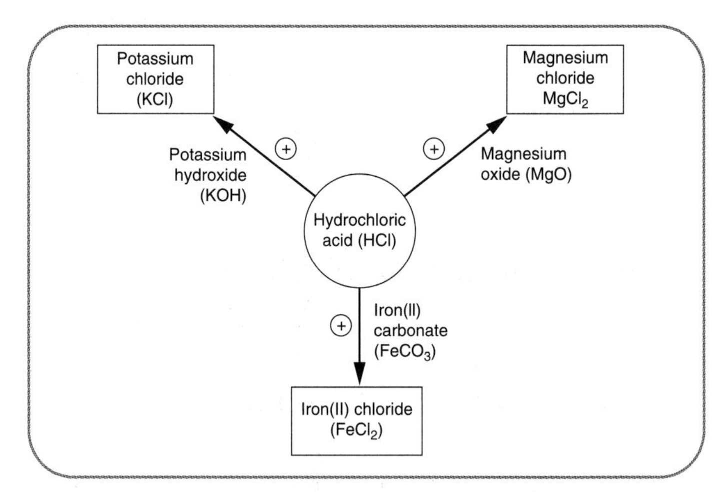

Figure 9.2 Salts of hydrochloric acid

Remember

The data booklet gives information on the solubility of salts.

Preparation of a soluble salt

1 Choose a suitable acid and an insoluble base (neutraliser).
2 Add an excess of the neutraliser to the acid.
3 Filter out the excess unreacted neutraliser.
4 Evaporate off the water, leaving the soluble salt.

For example, to make the soluble salt copper(II) sulphate, use copper(II) carbonate and sulphuric acid.

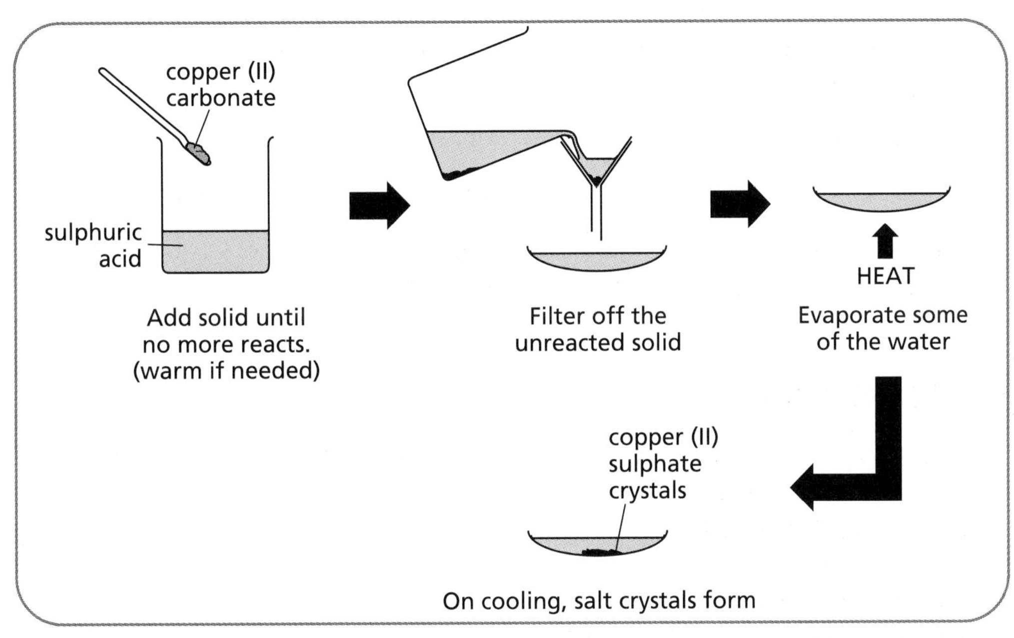

Figure 9.3 Making copper(II) sulphate

Preparation of an insoluble salt

1 Choose soluble compounds which contain the ions in the insoluble salt.
2 Make up solutions of each of these compounds.
3 Add the two solutions together; a precipitate of the insoluble salt will form .
4 Filter out the solid precipitate, which is the insoluble salt.

For example, to make the insoluble salt lead(II) iodide, use solutions of lead(II) nitrate and potassium iodide.

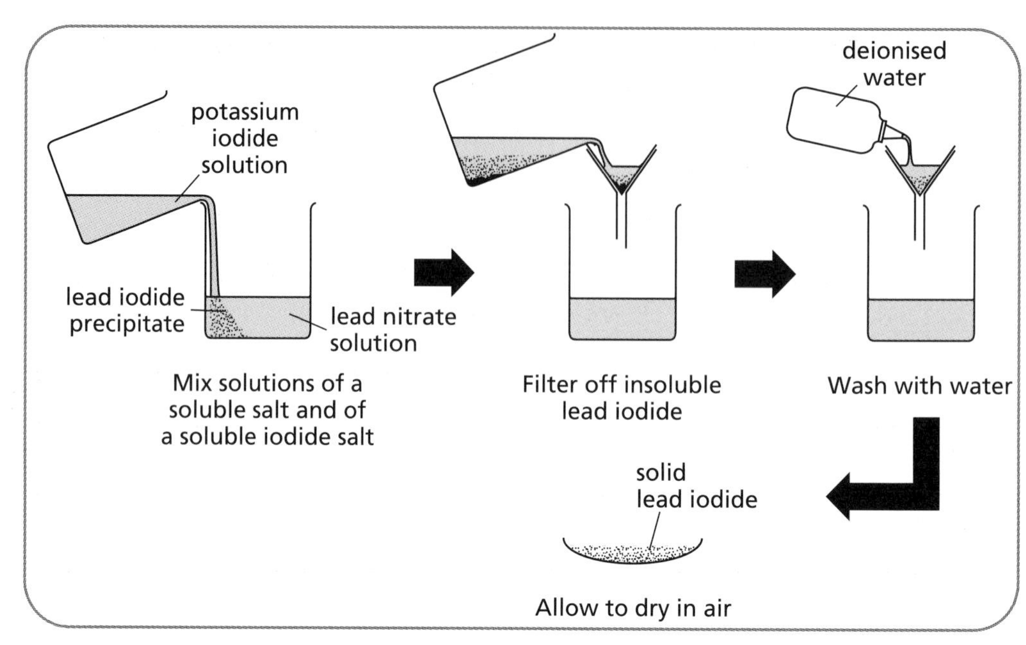

Figure 9.4 Making lead(II) iodide

A closer look at acid reactions

Neutralisation reactions

There are certain similarities in the reactions between all acids and all alkalis. Look at the equation for the neutralisation reaction between hydrochloric acid and sodium hydroxide shown below.

$$H^+(aq) + \cancel{Cl^-(aq)} + \cancel{Na^+(aq)} + OH^-(aq) \rightarrow \cancel{Na^+(aq)} + \cancel{Cl^-(aq)} + H_2O(l)$$

The $Na^+(aq)$ and $Cl^-(aq)$ ions remain unchanged by the reaction. These are called **spectator ions**. If they are removed the reaction equation becomes:

Equation To Remember

$$H^+(aq) + OH^-(aq) \rightarrow H_2O(l)$$

This is the same for all acids and alkalis. In these neutralisation reactions the hydroxide (OH^-) ions act as a **base** as they combine with the hydrogen (H^+) ions to form water.

The carbonate ion, in metal carbonates, can also act as a base in reactions with acids. Look at the equation for the neutralisation reaction between sodium carbonate solution and sulphuric acid.

$$2H^+(aq) + \cancel{SO_4^{2-}(aq)} + \cancel{2Na^+(aq)} + CO_3^{2-}(aq) \rightarrow \cancel{2Na^+(aq)} + \cancel{SO_4^{2-}(aq)} + H_2O(l) + CO_2(g)$$

In this reaction the spectator ions are $Na^+(aq)$ and the $SO_4^{2-}(aq)$. If they are removed the equation becomes:

$$2H^+(aq) + CO_3^{2-}(aq) \rightarrow H_2O(l) + CO_2(g)$$

Thus the carbonate (CO_3^{2-}) ions act as a base as they combine with the hydrogen (H^+) ions to form water and carbon dioxide.

Key Point

During all neutralisation reactions the hydrogen (H^+) ions of the acid react to make water.

Metal and acid reactions

The reaction of acids with metals is another important reaction of acids. In this reaction a salt and hydrogen gas are produced.

Equation To Remember

acid + **metal** → **salt** + **hydrogen**

For example, if magnesium is added to hydrochloric acid the salt magnesium chloride and hydrogen gas are formed.

$$2HCl + Mg \rightarrow MgCl_2 + H_2$$

hydrochloric acid + magnesium → magnesium chloride + hydrogen

The salt produced, magnesium chloride, is white so it dissolves forming a colourless solution. As the metal dissolves bubbles of hydrogen gas can be seen.

The reactions of acids and metals are dealt with in more detail in Chapter 11.

Reactions of acid rain

The reactions of acids with metal oxides, metal hydroxides, metal carbonates and metals help to explain some of the harmful effects of acid rain.

- Acid rain reacts with metal oxides in the soil and washes out essential minerals. This affects plant growth and damages life in rivers and lakes.

- Acid rain damages buildings made of limestone and marble as acids react with all metal carbonate rocks.
- Acid rain damages iron and steel objects like railings and car bodies, as acids react with metals.

Titrations

The amount of acid or alkali required for neutralisation depends on the **concentration** and the **volume** of the solutions used. To investigate a neutralisation reaction, chemists carry out a **titration**. This technique involves using special apparatus to measure the volumes of solutions accurately. The stages in a titration are outlined below

1. An exact volume of alkali is measured using a **pipette** and placed in a conical flask.
2. A suitable indicator is added to the alkali.
3. A **burette** is filled with the acid, and the starting volume is noted.
4. The acid is slowly added to the alkali until the indicator changes colour.

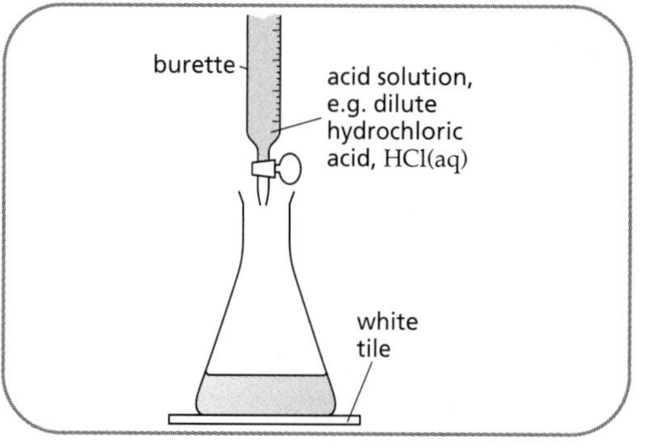

Figure 9.5 Slowly add acid from a burette, until the indicator changes colour

5. The final volume of acid is noted and recorded.
6. The titration is repeated to obtain accurate results.

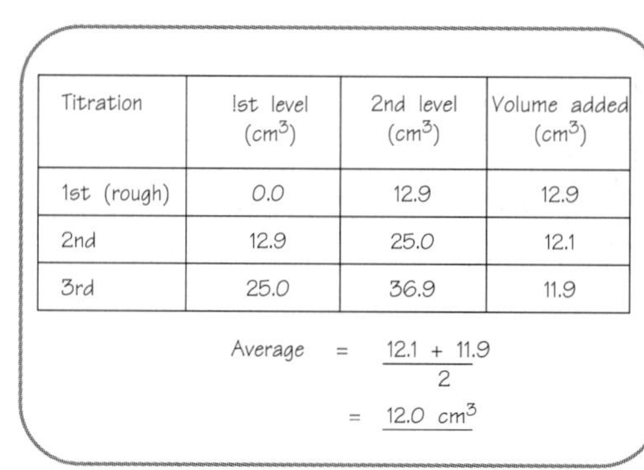

Titration	1st level (cm^3)	2nd level (cm^3)	Volume added (cm^3)
1st (rough)	0.0	12.9	12.9
2nd	12.9	25.0	12.1
3rd	25.0	36.9	11.9

$$\text{Average} = \frac{12.1 + 11.9}{2} = 12.0\ cm^3$$

Figure 9.6 Titration results

Calculations from titration results

The results of a titration can be used to calculate an unknown concentration of acid or alkali. The following equation can be used in these calculations.

Key Point

$$C_{acid} \cdot V_{acid} \cdot N_{H+} = C_{alkali} \cdot V_{alkali} \cdot N_{OH-}$$

where

C_{acid} = concentration of acid

V_{acid} = volume of acid

N_{H+} = number of H^+ ions in acid formula

C_{alkali} = concentration of alkali

V_{alkali} = volume of alkali

N_{OH-} = number of OH^- ions in alkali formula

Example

If 25.0 cm^3 of a solution of sodium hydroxide is exactly neutralised by 13.5 cm^3 of 0.6 mol/l sulphuric acid, what is the concentration of the alkali?

The known values are:

$C_{acid} = 0.6$ $C_{alkali} = ?$

$V_{acid} = 13.5$ $V_{alkali} = 25.0$

$N_{H+} = 2$ $N_{OH-} = 1$

Substitute these values in the equation:

$C_{acid} \cdot V_{acid} \cdot N_{H+} = C_{alkali} \cdot V_{alkali} \cdot N_{OH-}$

$0.6 \times 13.5 \times 2 = C_{alkali} \times 25.0 \times 1$

Simplify:

$16.2 = C_{alkali} \times 25.0$

Rearrange:

$C_{alkali} = \frac{16.2}{25.0} = 0.648$

$\therefore$ Concentration of alkali = 0.648 mol/l

Summary

- Neutralisation occurs when an acid reacts with a base to produce a salt and water.
- Metals oxides, metal hydroxides and metal carbonates are all examples of bases.
- An alkali is a base that is soluble in water.
- Acid + alkali → salt + water, e.g. $HNO_3 + NaOH \rightarrow NaNO_3 + H_2O$
- Acid + metal oxide → salt + water, e.g. $H_2SO_4 + MgO \rightarrow MgSO_4 + H_2O$
- Acid + metal carbonate → salt + water + carbon dioxide, e.g. $2HCl + CuCO_3 \rightarrow CuCl_2 + CO_2 + H_2O$
- During neutralisations the H^+ ions of the acid are removed and the pH rises towards 7.
- Acid + metal → salt + hydrogen, e.g. $2HCl + Zn \rightarrow ZnCl_2 + H_2$.
- Salts are formed when the H^+ ion of the acid is replaced with another positive ion.
- Examples of salts: $NaCl$, K_2SO_4, $Mg(NO_3)_2$ and $Cu_3(PO_4)_2$.
- The harmful affects of acid rain are due to the reactions of acids.
- A spectator ion is an ion in solution which is unchanged during a reaction.
- To make a soluble salt: add an insoluble base to an acid – filter out the excess base – then evaporate to leave salt.
- To make an insoluble salt: get solutions containing ions in the salt – add solutions to precipitate salt – filter out the salt.
- A titration is used to find unknown concentrations of acids or alkalis.
- From results of titration use: $C.V.N_{H+} = C.V.N_{OH-}$

Chapter 10

MAKING ELECTRICITY

Using batteries as a convenient source of electricity has become an important part of modern life. But how do batteries work and what happens to produce the electric current? This chapter looks at the chemistry involved in batteries and cells. It also introduces the idea of oxidation and reduction to help explain the chemical reactions involved.

Key Words

★ **battery** ★ **cell** ★ **displacement** ★ **electrochemical series** ★ **ion–electron equation** ★ **ion bridge** ★ **oxidation** ★ **rechargeable cell** ★ **redox reaction** ★ **reduction**

Electricity, batteries and cells

A **battery** is a piece of equipment that produces an electric current from a chemical reaction. Each battery, properly called a **cell**, consists of **two electrodes** in an **electrolyte**. The chemical reactions, which occur at the electrodes, involve the loss and gain of electrons. The electrons then flow through the wires from one electrode to the other, and the electrolyte completes the circuit.

One of the first cells to be made contained zinc and copper discs, separated by paper soaked in salt solution.

Note that the term 'battery' properly describes the set up where two or more cells are joined together. However, we often use it to describe the ordinary single cells we use every day.

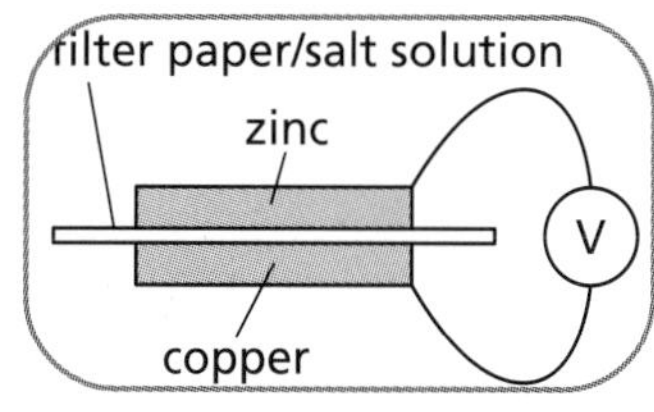

Figure 10.1 A simple cell

Different types of cells and batteries have been developed over the years and are still being developed. Two of the most common types of cell are shown on the next page.

The zinc–carbon cell is the common 'dry cell' battery used for torches, radios, etc. It is safe to use, as the chemical cannot leak. However it is not rechargeable and is thrown away when the chemicals are used up.

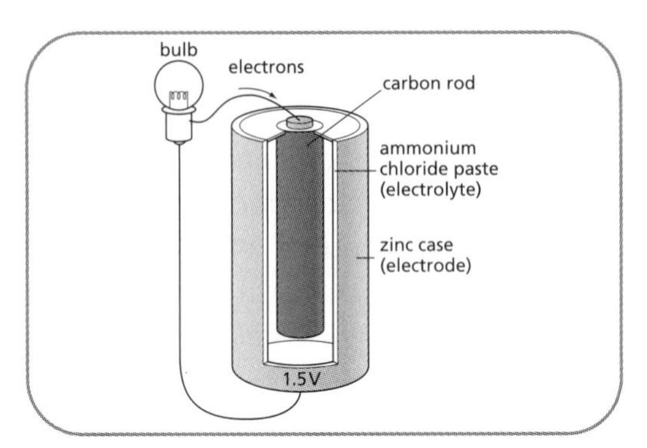

Figure 10.2 A zinc–carbon cell

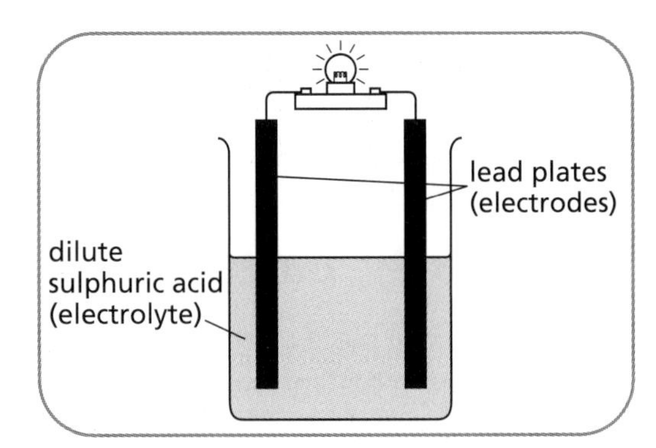

Figure 10.3 A lead–acid cell

The lead–acid cell found in car batteries is an example of a 'wet cell'. As the acid can leak it has to be used carefully. However, the cell is **rechargeable**, and can be made to work again by reversing the flow of electricity.

Reversing the flow of electricity in a rechargeable cell reverses the chemical reactions that produced the current in the first place.

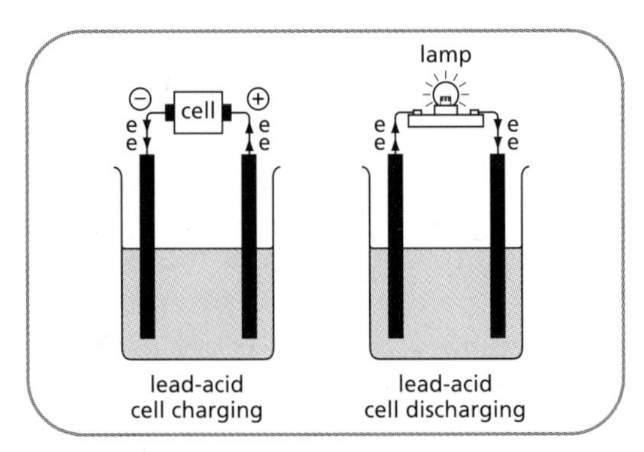

Figure 10.4 A lead–acid cell is rechargeable

Comparing batteries and mains electricity

Most of the electricity we use comes from power stations, rather than chemical cells. Each of these sources of electricity, mains and batteries, has certain advantages and disadvantages.

Mains electricity . . .	Cells and batteries . . .
◆ uses up finite resources of fossil fuels	◆ use up finite resources of metals
◆ is cheaper for general use	◆ are safer as they use a lower voltage
◆ gives more power	◆ are more portable and don't need cables

Cells and the electrochemical series

To investigate how cells work, a series of chemical cells can be set up using different metals.

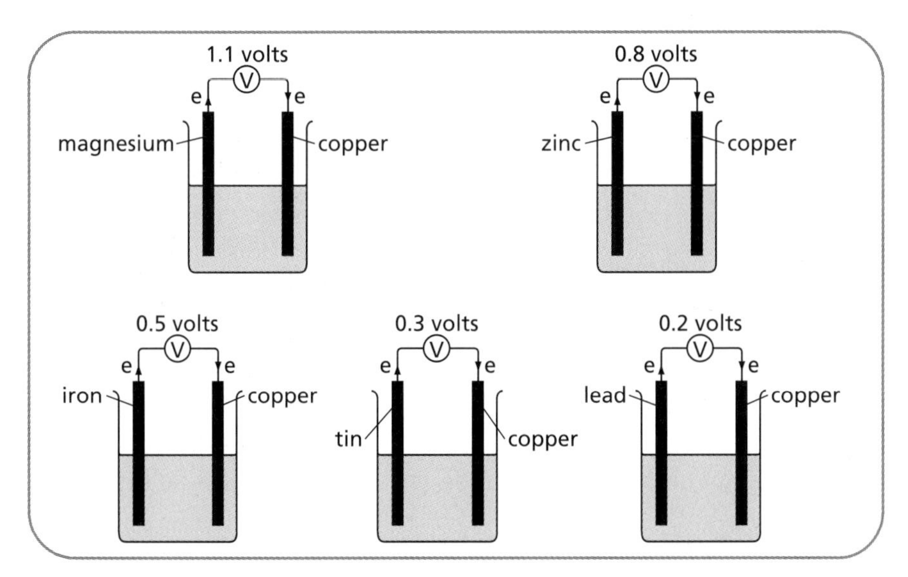

Figure 10.5 Investigating cells

The direction of electron flow and cell voltage tell you how easily the metals give away electrons. These results can be used to place the metals in an order called the **electrochemical series**.

The rules for the electrochemical series are as follows.

◆ Electrons flow from the metal higher up the series to the metal lower down.

◆ The metals furthest apart in the series produce the largest voltages and currents.

The electrochemical series for the metals used in these cells is:

magnesium

zinc

Iron

Tin

Lead

Copper

The full electrochemical series, which includes non-metals and group ions, can be found on page 7 of the *Chemistry Data Booklet*. This information is very useful when answering questions about cells and their reactions.

Cell reactions

To investigate the cell reactions, a standard cell can be set up, using metals in solutions of their own ions.

The separate solutions need to be joined by an **ion bridge** to complete the circuit. This is often just a filter paper soaked in an electrolyte solution, such as salt, so it conducts electricity.

The cell reactions that produce the electricity can be written as **ion–electron equations**. These equations show where the electrons are lost and gained.

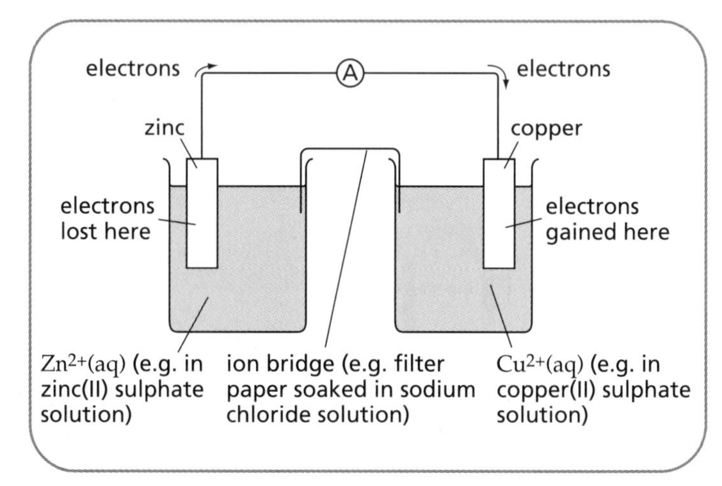

Figure 10.6 Standard zinc–copper cell

As electrons flow away from the zinc electrode the zinc atoms must be **losing electrons**.

$$Zn(s) \rightarrow Zn^{2+}(aq) + 2e^-$$

This means the zinc electrode gets lighter, as the zinc ions dissolve.

As electrons flow towards the copper electrode the copper ions must be gaining electrons.

$$Cu^{2+}(aq) + 2e^- \rightarrow Cu(s)$$

So the copper electrode gets heavier, as copper metal is formed on it.

Note that you do not need to work out the ion–electron equations for the cell. They can be found in the electrochemical series on page 7 of your *Chemistry Data Booklet*. As written they show electron gain: reverse them for the equation representing the loss of electrons.

Displacement reactions

The reactions that happen in a cell can be carried out directly by adding a metal to a solution of another metal's ions. This is called a **displacement** reaction. The rule is that metals higher up the electrochemical series can displace metals lower down the series.

For example, when zinc is added to copper(II) sulphate solution the following reaction occurs.

$$Zn(s) + Cu^{2+}(aq) + SO_4^{2-}(aq) \rightarrow Cu(s) + Zn^{2+}(aq) + SO_4^{2-}(aq)$$

zinc + copper(II) sulphate → copper + zinc(II) sulphate

The zinc solid dissolves as the zinc atoms lose electrons and form zinc ions.

$$Zn(s) \rightarrow Zn^{2+}(aq) + 2e^-$$

The blue colour of the copper(II) ions disappears, as the copper ions gain electrons and form the brown solid copper metal.

$$Cu^{2+}(aq) + 2e^- \rightarrow Cu(s)$$

The ion–electron equations are exactly the same as in the cell but this time we call it a displacement reaction. No electrical energy is released but heat energy is given out instead.

Note that the reactions of metals with acids, which is used to establish the position of hydrogen in the electrochemical series, is also a displacement reaction. The metals above hydrogen in the series will react with acids to displace hydrogen gas. The metals below hydrogen do not react with dilute acids

Cells without metals

Cell reactions always involve the loss and gain of electrons. However, they do not always have to include metals or metal ions. Look at the cell shown opposite. The electrodes are made of carbon, which is unreactive, and it is the ions in solution that react.

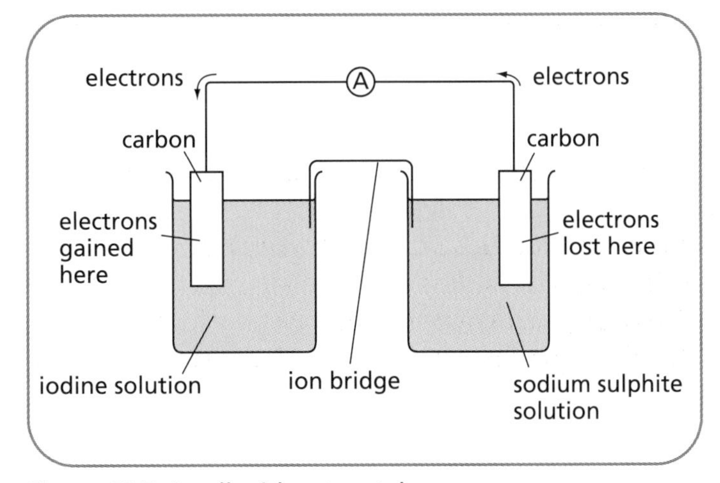

Figure 10.7 A cell without metals

Once again ion–electron equations can be written for the cell and the direction of electron flow tells you where electrons are lost and gained.

As the electrons flow away from the side with sulphite solution, these ions must lose electrons.

$$SO_3^{2-}(aq) + H_2O(l) \rightarrow SO_4^{2-}(aq) + 2H^+(aq) + 2e^-$$

As the electrons flow towards the side with iodine solution, the iodine molecules must gain electrons.

$$I_2(aq) + 2e^- \rightarrow 2I^-(aq)$$

Redox reactions

The reactions that occur in cells and in the displacement of metal ions are called **redox reactions**. These are reactions that involve the loss and gain of electrons.

Key Points

The loss of electrons by an atom or ion is called **oxidation**.

The gain of electrons by an atom or ion is called **reduction**.

Oxidation and reduction always occur together and so the overall change is called a redox reaction.

Remember

You can remember the meaning of oxidation and reduction by the phrase 'OIL RIG'.

Oxidation	**R**eduction
Is	**I**s
Loss of e^-	**G**ain of e^-

The **electrochemical series** in the data booklet contains many examples of ion–electron equations for oxidation and reduction. Note that the equations are written as reduction reactions, from left to right. If you reverse the equations, and go from right to left, they represent oxidation reactions.

Example

Some examples of oxidation and reduction ion–electron equations are shown below.

Reduction:	Oxidation:
$Al^{3+}(aq) + 3e^- \rightarrow Al(s)$	$Mg(s) \rightarrow Mg^{2+}(aq) + 2e^-$
$Pb^{2+}(aq) + 2e^- \rightarrow Pb(s)$	$Ag(s) \rightarrow Ag^+(aq) + e^-$
$Br_2(l) + 2e^- \rightarrow 2Br^-(aq)$	$2Cl^-(aq) \rightarrow Cl_2(g) + 2e^-$

In general, metal atoms and non-metal ions are oxidised and lose electrons. Metal ions and non-metal atoms are usually reduced and gain electrons.

The reaction that occurs between a metal and an acid is another example of a redox reaction that can be described by ion–electron equations.

For example, the reaction between iron and hydrochloric acid can be described as follows.

$$Fe(s) + 2H^+(aq) + 2Cl^-(aq) \rightarrow Fe^{2+}(aq) + 2Cl^-(aq) + H_2(g)$$

iron + hydrochloric acid → iron(II) chloride + hydrogen

The ion–electron equations are shown below.

oxidation: $Fe(s) \rightarrow Fe^{2+}(aq) + 2e^-$

reduction: $2H^+(aq) + 2e^- \rightarrow H_2(g)$

In this reaction the chloride ion (Cl^-) is a spectator ion as it is unchanged.

Hints and Tips

Use of ion charges

When dealing with redox reactions it is often useful to include ion charges in the formula. For example:

$$Cu^{2+}SO_4^{2-}(s) \text{ or } Cu^{2+}(aq) + SO_4^{2-}(aq)$$

$$Mg^{2+}(Cl^-)_2 (s) \text{ or } Mg^{2+}(aq) + 2Cl^-(aq)$$

$$(Al^{3+})_2(SO_4^{2-})_3(s) \text{ or } 2Al^{3+}(aq) + 3SO_4^{2-} (aq)$$

You can also use ion–electron equations to describe the **electrolysis** process. Look at the diagram below which shows what happens when copper chloride is electrolysed.

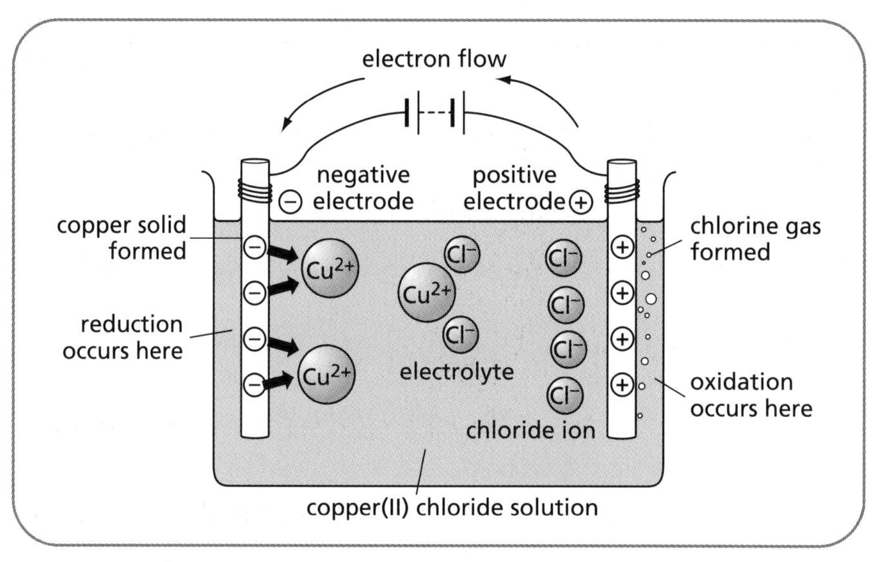

Figure 10.8 Electrolysis, a redox reaction

The ion-electron equations for this electrolysis are shown below.

oxidation:	$2Cl^-(aq) \rightarrow Cl_2(g) + 2e^-$	at the positive electrode
reduction:	$Cu^{2+}(aq) + 2e^- \rightarrow Cu(s)$	at the negative electrode

The direction of electron flow tells you where oxidation and reduction occur. During electrolysis oxidation always occurs at the positive electrode while reduction occurs at the negative electrode.

Summary

- A cell or battery produces an electric current from a chemical reaction.
- Electricity passing through a metal involves a flow of electrons.
- Electricity passing through a solution involves the movement of ions.
- All cells consist of two electrodes with an electrolyte between to complete the circuit.
- Zinc–carbon cells have to be thrown away when the chemicals are used up.
- Lead–acid cells are rechargeable by reversing the flow of electricity.
- The main advantages of using batteries and cells are that they are safe and portable.
- The main advantages of using mains electricity is it that it is cheaper and higher power.
- Joining two different metals in a salt solution makes a simple cell.
- The electrons flow from the metal higher up the electrochemical series to the metal lower down the series.
- The largest voltages and currents are obtained when the metals are furthest apart in the electrochemical series.
- In a cell with two separate solutions the ion bridge completes the circuit.
- Metals higher up the electrochemical series can displace metals lower down the series.
- The reactions of metals with acid places hydrogen in between lead and copper in the electrochemical series.
- In redox reactions, oxidation is the loss of electrons and reduction is the gain of electrons.
- Examples of redox reactions include: cells, displacement, metals and acids and electrolysis.
- Ion–electron equations showing electrons lost and gained, are found page 7 of the data booklet.

Chapter 11

THE REACTIONS OF METALS

About three quarters of the elements in the Periodic Table are metals, and they have a wide variety of properties and uses. But how are metals different from non-metals, and what are the main similarities and differences between metals? This chapter looks at the reactions, properties and methods of extraction of different metals and alloys. It also introduces more calculations involving the mole.

Key Words

★ **alloy** ★ **blast furnace** ★ **decomposition** ★ **ductile**
★ **empirical formula** ★ **malleable** ★ **native metal** ★ **ore**
★ **reactivity series** ★ **recycle** ★ **salt**

Properties and uses of metals

Metals are amongst our most valuable resources as they have a wide variety of properties and uses.

Key Points

The main physical properties of metals are:

- **Hard**, **strong**, **flexible** solids (only mercury is a liquid at room temperature).
- **Shiny** (when polished).
- Good **conductors** of **heat** (allow heat to pass through them).
- Good **conductors** of **electricity** (allow electricity to pass through them).
- **High density** (high mass per unit volume, g/cm^3).
- **Malleable** (able to be hammered into shape).
- **Ductile** (able to be drawn into wires).

In most cases the uses of metals are linked to their properties. Some examples of metals, properties and uses are shown in the table.

Properties and uses of some metals

Metal	Property	Used for making . . .
copper	a good conductor of electricity	electrical cables and circuits
iron	hard and strong	bridges and buildings
aluminium	a good conductor of heat	pots, pans and cooking foil
gold and silver	malleable and ductile	intricate jewellery

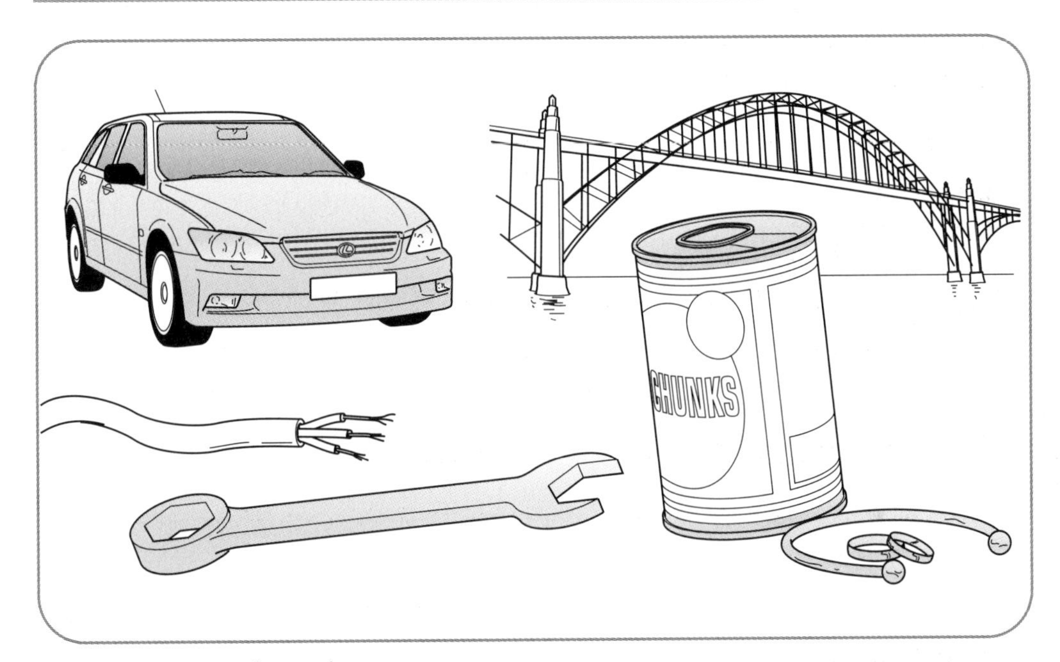

Figure 11.1 Uses of metal

Recycling metals

Metals are a **finite** resource, and there is only a limited supply of each metal. The demand for certain metals, like copper, tin and lead, is so great that we are using up our resources very quickly. We must therefore learn to **recycle** more metals, as by reusing them we will save our precious reserves for the future.

Alloys

A mixture of a metals with other elements is called an **alloy**. Making alloys produces a range of materials with improved properties. This is important as it makes the metals more useful.

Some common alloys

Alloy	Mixture of metals	Uses
steel	iron and carbon	girders for buildings
solder	tin and lead	connecting electrical circuits
brass	copper and zinc	ornaments, name plates, etc.

Reactions of metals

The most important chemical reactions of metals are with oxygen, water and acids. These are all common substances, found in nature and in the home. Although the chemical reactions of metals are similar, they do not all react in the same way.

Metals and oxygen

Most metals react with oxygen forming a metal oxide. Metals like potassium, sodium and lithium, which look shiny when cut, dull quickly in air as the surface of the metals react with oxygen. Some metals, like calcium, magnesium and aluminium burn brightly if heated in oxygen.

The general equation for the reaction of metals with oxygen is shown below, along with two examples for particular metals.

Equation To Learn

metal + **oxygen** → **metal oxide**

When copper is heated it turns black as copper(II) oxide forms on its surface.

$$2Cu + O_2 \rightarrow 2CuO$$

copper + oxygen → copper(II) oxide

When magnesium is heated in air it burns brightly, forming a white solid that is magnesium oxide.

$$2Mg + O_2 \rightarrow 2MgO$$

magnesium + oxygen → magnesium oxide

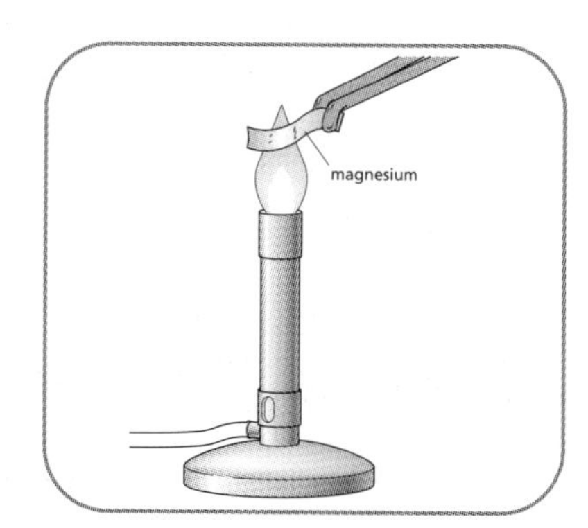

Figure 11.2 Magnesium burning

Metals and water

Some metals like potassium, sodium, lithium and calcium react quickly with cold water. These reactions produce hydrogen gas and an alkaline solution. The general equation for the reactions of metals with water is shown below.

Equation To Learn

metal + water → metal hydroxide + hydrogen

Example

Sodium reacts violently with water producing enough heat to melt the metal. Hydrogen gas is given off as alkaline sodium hydroxide solution is formed.

$$2Na + 2H_2O \rightarrow 2NaOH + H_2$$

sodium + water → sodium Hydroxide + hydrogen

Potassium reacts so violently with water that the hydrogen gas formed bursts into flames. An alkaline solution of potassium hydroxide is also formed.

$$2K + 2H_2O \rightarrow 2KOH + H_2$$

potassium + water → potassium Hydroxide + hydrogen

Metals like lithium, sodium and potassium are stored under oil, as they react with water vapour in the air.

Metals and acid

Metals like magnesium, zinc and iron, which do not react quickly with cold water, react with dilute acids. These reactions produce hydrogen gas and a new substance called a **salt**. The general equation for the reactions of metals with acids is shown below.

Equation To Learn

metal + acid → salt + hydrogen

Example

Zinc reacts quickly with hydrochloric acid forming the soluble salt called zinc(II) chloride and hydrogen gas.

$$Zn + 2HCl \rightarrow ZnCl_2 + H_2$$

zinc + hydrochloric acid → zinc(II) chloride + hydrogen

Magnesium reacts quickly with nitric acid forming soluble salt called magnesium nitrate and hydrogen gas.

$$Mg + 2HNO_3 \rightarrow Zn(NO_3)_2 + H_2$$

magnesium + nitric acid → magnesium nitrate + hydrogen

The reactions of metals, acids and salts are dealt with in more detail in Chapter 9.

Reactivity series

Experiments with metals can be used to construct a reactivity series. The example shown on the next page lists the metals in order of decreasing reactivity. A summary of the metals' main reactions is also included in the table.

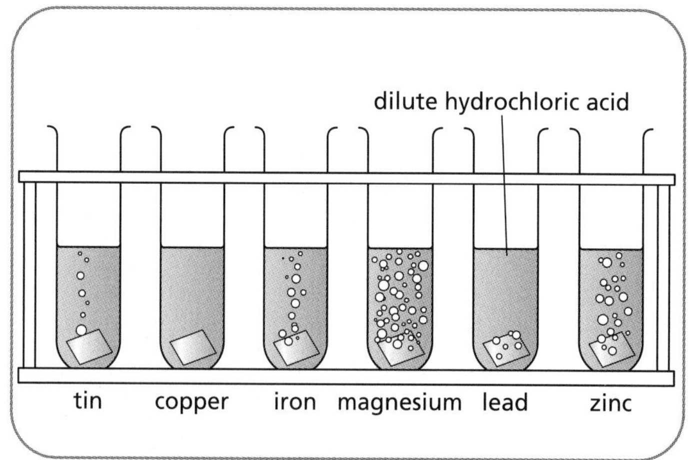

Figure 11.3 Experiments can be used to form a reactivity series for metals

	Reactivity series	Reaction with water	Reaction with dilute acids	Reaction with oxygen
↑	Potassium	Metals which react quickly with cold water forming a metal hydroxide + hydrogen	Metals which react with dilute acids forming hydrogen + salt	Metals which react with oxygen
	Sodium			
	Lithium			
R	Calcium			
E	Magnesium	Metals which do *not* react quickly with cold water		
A	Aluminium			
C	Zinc			
T	Iron			
I	Nickel			
V	Tin			
I	Lead			
T	Copper		Metals which do *not* react with dilute acids	
Y	Mercury			
	Silver			
	Gold			Metals which do *not* react with oxygen
	Platinum			

Hints and Tips

You can use the following to remember the Reactivity Series.

Peter's	**P**otassium
School	**S**odium
Council	**C**alcium
Meeting	**M**agnesium
At	**A**luminium
Zoo	**Z**inc
Induced	**I**ron
Night	**N**ickel
Time	**T**in
Laughter	**L**ead
Causing	**C**opper
Many	**M**ercury
Sleepy	**S**ilver
Grumpy	**G**old
Pupils	**P**latinum

Whatever the reaction, a metal higher up the **reactivity series** will generally react faster than a metal lower down the series. Therefore the series is useful for making predictions about other reactions of metals.

For example, displacement reactions generally occur when the metal being added is higher up the reactivity series than the metal ion in solution. So iron can displace copper from copper sulphate, but copper cannot displace iron from its compounds.

Note that if you forget the reactivity series you can use the electrochemical series on page 7 of the *Chemistry Data Booklet*, as they are very similar.

See Chapter 10 for more about displacement reactions and the electrochemical series.

Sources of metals

Only a few metals like **gold**, **silver** and **platinum** can be found uncombined in the ground. These are called **native metals**. They are found as elements as they are very unreactive.

Most metals, however, react with the substances around them and are therefore found in **compounds**. Rocks which contain metal compounds and are used as a source of the metal are called **ores**. Ores are usually mixtures of substances but commonly contain oxides, sulphides or carbonates of metals. Some examples of common metal ores are given below.

- Haematite, an ore of iron, is mainly iron oxide.
- Bauxite, an ore of aluminium, is mainly aluminium oxide.
- Galena, an ore of lead, is mainly lead sulphide.
- Malachite, an ore of copper, is mainly copper carbonate.

To **extract** a metal, its ore has to be broken down or **decomposed**. There three main methods used to decompose metal compounds.

1 Heat alone.
2 Heat with carbon.
3 Electrolysis.

The method chosen depends on the reactivity of the metal. The most reactive metals form the most stable compounds which are hardest to break down.

Note that the extraction of a metal is sometimes described as a **reduction** reaction, as the metal ion in the ore is reduced. For example, in the extraction of iron the following reaction occurs:

$$Fe^{3+} + 3e^- \rightarrow Fe$$

This is reduction as the iron(III) ions gain electrons to form iron atoms.

Extraction by heat alone

Using heat alone is the simplest and usually the cheapest method that can be used to extract a metal. Only the **least reactive** metal ores can be broken down by this method. The metals mercury and silver are commonly extracted by heat alone.

For example, if mercury(II) oxide is heated it breaks down to release silvery droplets of mercury and oxygen gas.

$$2HgO(s) \rightarrow 2Hg(l) + O_2(g)$$

mercury(II) oxide → mercury + oxygen

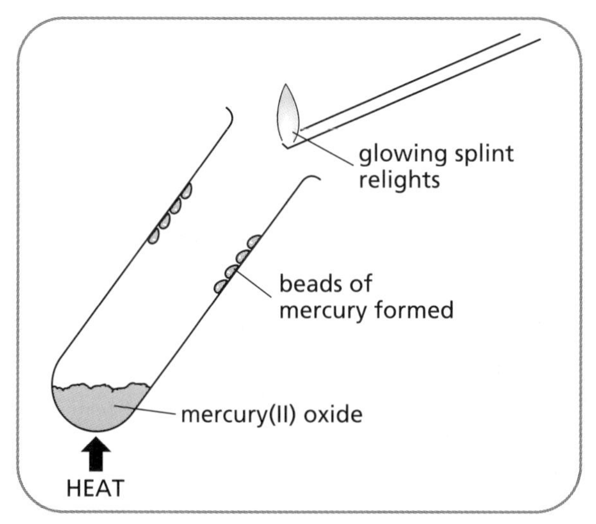

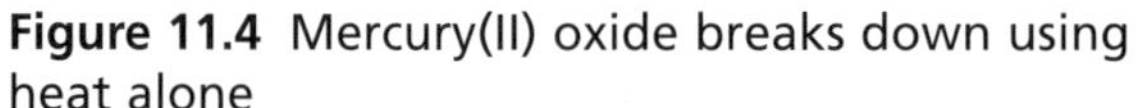
Figure 11.4 Mercury(II) oxide breaks down using heat alone

Extraction by heating with carbon

Metal oxides which cannot be reduced by heat alone can often be broken down by heating with carbon. Zinc, iron, nickel, tin, lead and copper, which are found in the middle of the reactivity series, are usually extracted by heating with carbon.

For example, if copper(II) oxide is heated with carbon powder, brown copper metal is released and carbon dioxide gas is produced.

$$2CuO(s) + C(s) \rightarrow 2Cu(s) + CO_2(g)$$

copper(II) oxide + carbon → copper + carbon dioxide

This is an example of a redox reaction. The carbon acts as a reducing agent, as it takes the oxygen away from the metal and forms carbon dioxide.

The reaction can also be described as a displacement reaction. Carbon is more reactive than these metals and displaces them from their ores.

The blast furnace

Iron is extracted from its ore, iron(III) oxide, in a **blast furnace**. The raw materials for the process, **iron ore**, **limestone** and **coke** (an impure form of carbon), are fed into the top of the furnace while hot air is blown in at the bottom. The main chemical reactions that occur are as follows.

1 The carbon (coke) burns in the air to produce carbon dioxide gas and heat.

$$C(s) + O_2(g) \rightarrow CO_2(g)$$

carbon + oxygen → carbon dioxide

2 Due to the high temperatures the carbon dioxide reacts with more carbon to form carbon monoxide gas.

$$CO_2(g) + C(s) \rightarrow 2CO(g)$$

carbon dioxide + carbon → carbon monoxide

3 The carbon monoxide gas reacts with the iron(III) oxide, removing the oxygen to form carbon dioxide and release the metal.

$$3CO(g) + Fe_2O_3(s) \rightarrow 2Fe(s) + 3CO_2(s)$$

carbon monoxide + iron(III) oxide → iron + carbon dioxide

Extraction by Electrolysis

The electrolysis of any molten ore can break down the compound and release the metal. However, as electrolysis needs large quantities of electricity, it is very expensive. So it is only used when no other method will work.

Only the most reactive metals like potassium, sodium, lithium, calcium, magnesium and aluminium are extracted by electrolysis.

Aluminium is extracted by the electrolysis of molten aluminium oxide. This is an electrolyte as it contains ions. The aluminium ions are positive (Al^{3+}) so the metal is formed at the negative electrode while oxygen is formed at the positive electrode. During electrolysis the electrical energy is used to decompose the metal compound.

See Chapter 7 for more about electrolysis.

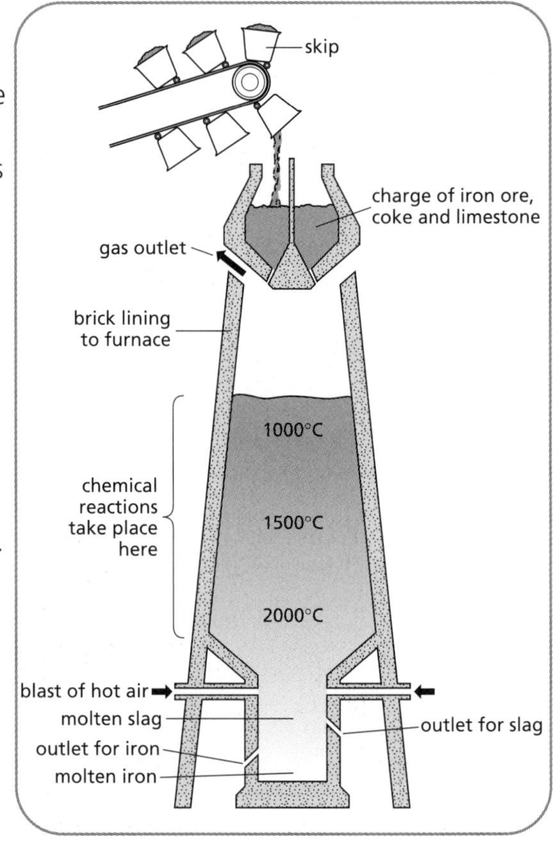

Figure 11.5 The blast furnace

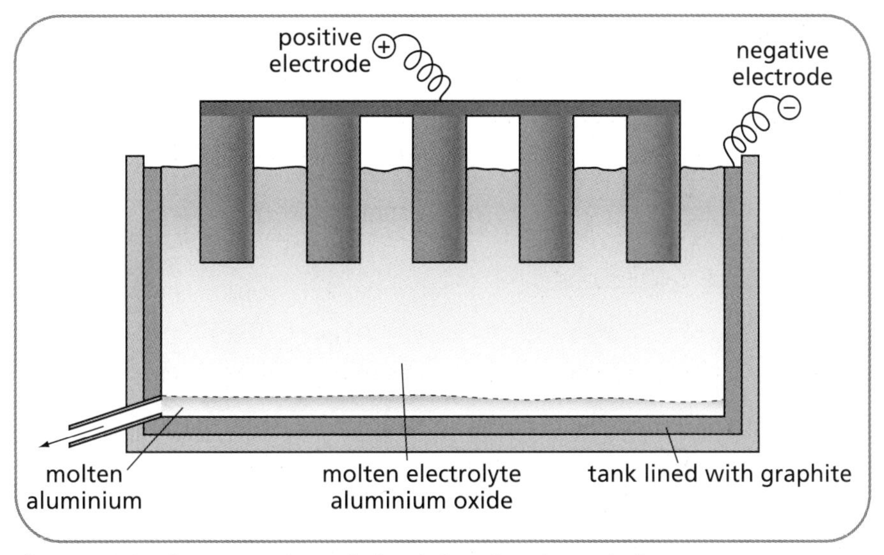

Figure 11.6 The extraction of aluminium by electrolysis

Discovery of metals

The dates of discovery of metals are linked to their reactivity.

- Gold and silver were discovered first as they could be found **uncombined** in the earth's crust.
- Copper and tin were next, in the **Bronze Age**, as their ores are easily broken down.
- Iron and the **Iron Age** came later, as higher temperatures were needed to reduce its ore.
- As **electricity** is needed to produce aluminium, magnesium and sodium, they were not discovered until the nineteenth century.

Figure 11.7 The least reactive metals were the first to be extracted by early humans

More calculations using moles

Relative atomic masses, formula masses and moles can be used in calculations of percentage composition, empirical formula and the mass of reactant or product in a reaction. Students often find questions involving calculations and the mole difficult. However, with a bit of practice you can improve your skills in this area. Try to become familiar with the different types of question and how each can be tackled.

Figure 11.8 Practice calculations

Percentage composition

To calculate the percentage, by mass, of an element in a compound, first work out its formula mass, and then use the equation below to calculate the percentage composition.

$$\textbf{\% of element} = \frac{\textbf{mass of element}}{\textbf{formula mass}} \times \textbf{100}$$

Example 1

Calcium nitrate could be used as a fertiliser as it contains nitrogen. What is the percentage, by mass, of the element nitrogen in calcium nitrate?

First work out the formula mass:

$Ca(NO_3)_2$

- O: $16 \times 6 = 96$
- N: $14 \times 2 = 28$
- Ca: $40 \times 1 = 40$

∴ Formula mass = 164

Then substitute values from the question into the equation.

% nitrogen = $\frac{28}{164} \times 100$ (there are two nitrogen atoms, so mass = 28)

∴ **% mass of nitrogen = 17.07%**

Empirical formula

The **empirical formula** of a compound is the simplest ratio of the elements in the compound. It can be worked out from the percentage composition or the masses of the elements that are combined. One way of calculating an empirical formula is to use the steps listed in the example below.

Example 2

A chemist found by analysis that 4.48 g of iron was combined with 1.92 g of oxygen. What is the formula of this oxide of iron?

This can be answered in a table.

Element	Mass of element (g)	Mass of 1 mole (g)	Number of moles present	Whole number ratio in moles
Fe	4.48	56	$\frac{4.48}{56} = 0.08$	$\frac{0.08}{0.08} = 1 = 2$
O	1.92	16	$\frac{1.92}{16} = 0.12$	$\frac{0.12}{0.08} = 1.5 = 3$
Notes:			*(divide mass of element by mass of one mole)*	*(divide each by the smallest number of moles)*

∴ **The empirical formula is Fe_2O_3**

Calculations using equations

Sometimes you are asked to find the **mass** of a **reactant** used or **product** formed in a chemical reaction. If you recognise a question involves this type of calculation, use the stages outlined in the box on the next page.

Hints and Tips

Steps involved in calculating amounts of reactant or product:

1 **E**quation, balanced for the reaction involved in the question.
2 **M**ole ratio, of the substances in the question.
3 **U**nits, change the moles to grams if necessary.
4 **S**olve, for the amounts in the question.

Use **'EMUS'** to remember the stages.

Example 3

What mass of iron would be produced by the reduction of 20 kg of iron(III) oxide using carbon as the reducing agent?

Step 1 The balanced chemical equation for the reaction.

$$2Fe_2O_3 + 3C \rightarrow 4Fe + 3CO_2$$

Step 2 Find the mole ratio of the substances in the question.

2 moles of Fe_2O_3 $\rightarrow$ 4 moles of Fe

Step 3 Change the units from moles to grams.

320 g of Fe_2O_3 $\rightarrow$ 224 g of Fe

Formula mass of Fe_2O_3
= (56 × 2) + (16 × 3)
= 160
1 mole = 160 g
∴ 2 moles = 320 g

1 mole of Fe
= 56 g
∴ 4 moles = 224 g

Step 4 Solve for the amounts in the question.

That is, 20 kg or 20 000 g of iron(III) oxide.
(as 1 kg = 1000 g)

1 g of Fe_2O_3 $\rightarrow$ 224/320 g of Fe

$$\therefore 20\,000\text{ g of } Fe_2O_3 \rightarrow 20\,000 \times \frac{224}{320} = 14\,000\text{ g of Fe}$$

∴ 20 kg of Fe_2O_3 $\rightarrow$ 14 kg of Fe

Summary

- Metals are generally strong, hard, dense, malleable, ductile, good conductors of heat and good conductors of electricity.
- We need to recycle (reuse) metals, as they are finite and could be used up.
- An alloy is a mixture of metals with other elements, e.g. steel and solder.
- The order of reactivity is
  K > Na > Li > Ca > Mg > Al > Zn > Fe > Ni > Sn > Pb > Cu > Hg > Ag > Au > Pt.
- Potassium, sodium and lithium are stored under oil, as they react quickly with air.
- Potassium to silver react with oxygen → metal oxide.
- Potassium to calcium react quickly with cold water → metal hydroxide + hydrogen.
- Potassium to lead react with dilute acids → salt + hydrogen.
- The electrochemical series in the data book is similar to the reactivity series.
- A metal higher up the reactivity series can displace a metal lower down the series from its compounds.
- The least reactive metals, silver, gold and platinum, are found native, uncombined, as they are very unreactive.
- An ore is a rock that is used as a source of the metal.
- The fairly unreactive metals, mercury and silver, are extracted by heat alone.
- The medium reactive metals, copper → zinc, are extracted by heating with carbon.
- The most reactive metals, aluminium → potassium, are extracted using electricity.
- Iron is extracted from iron ore using coke (carbon) in a blast furnace.
- In a compound: $\% \text{ element} = \frac{\text{mass of element}}{\text{formula of mass}} \times 100$
- The empirical formula of a compound is the simplest ratio of its elements.
- A balanced chemical equation can be used to calculate the amount of reactant or product involved in a reaction.

Chapter 12

CORROSION

The corrosion of metals, and in particular the rusting of iron, costs this country millions of pounds every year. But why do metals corrode and how can we stop it happening? This chapter looks at the chemical reactions involved in corrosion and rusting. It also investigates a variety of ways of preventing corrosion.

Key Words

★ **corrosion** ★ **electroplating** ★ **galvanising** ★ **ion-electron equations**
★ **oxidation** ★ **redox reaction** ★ **reduction** ★ **rusting**
★ **sacrificial protection** ★ **surface barrier** ★ **tin-plating**

Corrosion

When exposed to air most metals form a dull surface coating, which usually consists of the oxide of the metal. For example, if a piece of sodium is scratched the shiny metal surface can be seen underneath. However, this quickly becomes dull as the metal corrodes and forms sodium oxide.

$$4Na(s) + O_2(g) \rightarrow 2Na_2O(s)$$

sodium + oxygen → sodium oxide

Corrosion is the reaction that forms a compound on the surface of a metal. It is an example of an **oxidation** reaction as during corrosion the metal atoms lose electrons and form positive metal ions. Aluminium atoms lose three electrons when they corrode.

$$Al \rightarrow Al^{3+} + 3e^-$$

Corrosion is usually a slow process, however the more reactive the metal the faster it will corrode. The surface of a fresh piece of zinc will take several hours to corrode, while a scratched piece of sodium will become dull in seconds.

Rusting

The corrosion of iron and steel is called **rusting**. As these are the most widely used metals in the world, rusting is a common problem everywhere.

Iron and steel rust when exposed to air and water. You can investigate the conditions needed for iron to rust by setting up a series of experiments

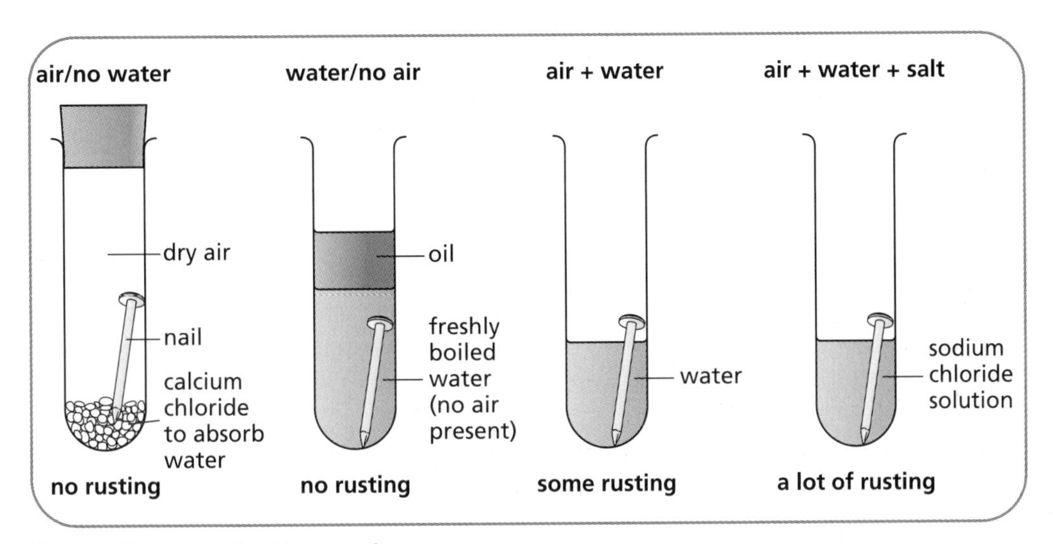

Figure 12.1 Investigating rusting

There are two main conclusions from these experiments.

1 Both **air** and **water** must be present for iron to rust.

2 The presence of **salt** dissolved in the water increases rusting.

Further experiments have shown that rust is a complex mixture of iron compounds, formed by reaction of iron with oxygen and water. The main compound in rust is iron(III) hydroxide.

Explaining rusting

As rusting is a **redox reaction** it can be investigated by setting up a cell. In the cell shown below, ferroxyl indicator is used to test for the presence of $Fe^{2+}(aq)$ and $OH^-(aq)$ ions

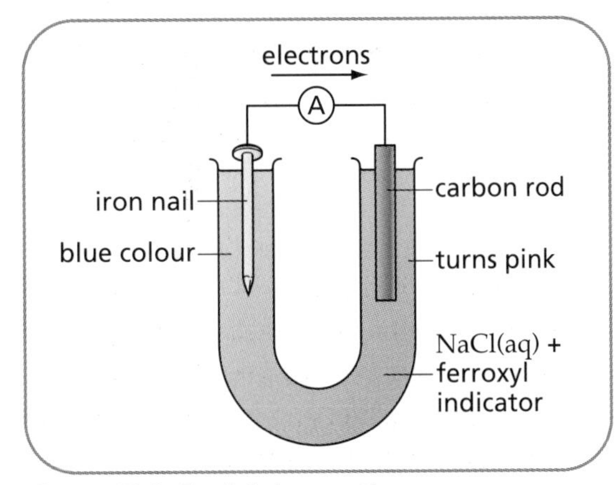

Figure 12.2 Explaining rusting

Ferroxyl indicator changes colour . . .
from green to blue if $Fe^{2+}(aq)$ ions are present
from green to pink if $OH^{-}(aq)$ ions are present.

The direction of electron flow tells you that oxidation occurs at the iron nail and reduction occurs at the carbon rod.

The ferroxyl indicator changing colour from green to blue around the iron tells you that $Fe^{2+}(aq)$ ions must be formed by the oxidation.

$$Fe(s) \rightarrow Fe^{2+}(aq) + 2e^{-}$$

Further experiments show that a two-stage oxidation occurs, and the $Fe^{2+}(aq)$ ions lose another electron to form $Fe^{3+}(aq)$ ions.

$$Fe^{2+}(aq) \rightarrow Fe^{3+}(aq) + e^{-}$$

The ferroxyl indicator changing colour from green to pink around the carbon tells you that $OH^{-}(aq)$ ions are formed by the reduction reaction. The electrons lost by the iron must be gained by the water and oxygen.

$$2H_2O(l) + O_2(g) + 4e^{-} \rightarrow 4OH^{-}(aq)$$

As before you do not need to remember the ion–electron equations for rusting. They can be found in the electrochemical series on page 7 of the *Chemistry Data Booklet*.

Electrolytes and rusting

As rusting is a redox reaction it needs the presence of an electrolyte solution. This is similar to the need for an electrolyte in a cell. Pure water normally contains a few ions from dissolved carbon dioxide. Adding an electrolyte to water will increase the concentration of the ions, and so speed up rusting. This explains why the use of salt on icy roads speeds up the rusting of cars. The salt dissolves in the rainwater and acts as an extra electrolyte.

Preventing corrosion

The rust formed on iron and steel is brittle and breaks away easily. This allows more rusting to occur, which eventually leads to the destruction of the metal object. The corrosion of iron and steel costs this country millions of pounds each year. Understanding how rusting occurs helps us find ways of preventing it. There are two main ways of protecting iron and steel from rusting.

- **Sacrificial protection**: Attaching iron to a metal higher up the electrochemical series, to prevent oxidation of the iron.
- **Surface barrier protection**: Coating the iron to keep out air and water to stop the iron rusting.

The simplest way of preventing rusting is a **surface barrier**. This keeps out oxygen and water so the iron cannot rust. There are several different surface barriers that can be used to protect iron and steel.

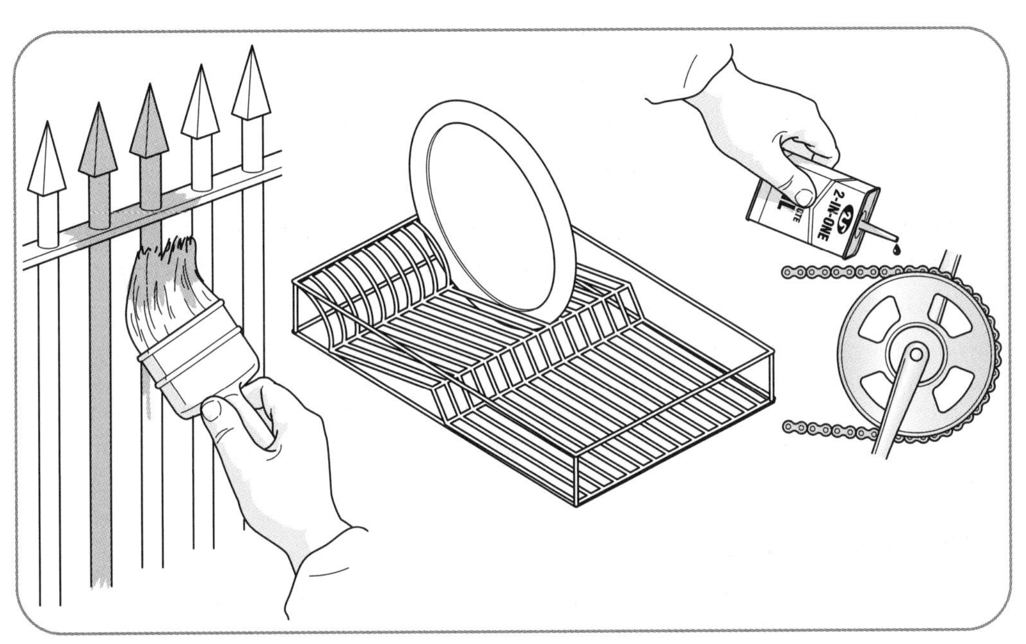

Figure 12.3 Preventing corrosion using a physical barrier

- Painting is used to protect many everyday steel objects like cars, bikes, bridges, etc.
- Greasing prevents the corrosion of tools and machinery.
- Plastic coatings are used for kitchen utensils and wire netting.
- Tin-plating is used for food containers.
- Galvanising, coating in zinc, is used for tanks, buckets and roofing nails.
- Electroplating is used for jewellery, cutlery and bathroom fitments.

Understanding sacrificial protection

It was discovered that iron and steel could be protected from rusting by attaching them to a metal higher up the electrochemical series (a more reactive metal), like magnesium or zinc. This unusual method of preventing rusting is useful where surface barriers are difficult to apply. It is called **sacrificial protection** and some examples of where it is used are described below.

- Bags of magnesium are attached at regular intervals to underground gas pipes.
- Iron ships often have blocks of zinc bolted to their hulls.
- Blocks of magnesium metal are attached to oil rigs in the North Sea.

The method works because the more reactive metal gives electrons to the iron. This makes it more difficult for the iron to lose electrons itself, and so it doesn't rust. Sacrificial protection works like a chemical cell, in which the magnesium or zinc is oxidised while the iron gains electrons. It is called **sacrificial protection** as the more reactive metal corrodes and is therefore sacrificed to save the iron.

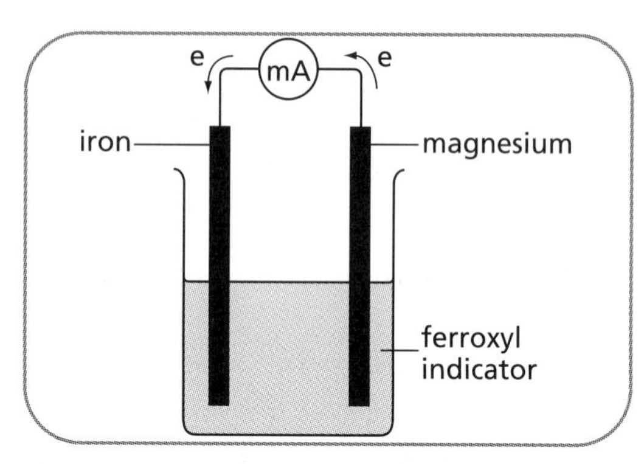

Figure 12.4 The magnesium corrodes and is sacrificed

The main disadvantage of sacrificial protection is that the more reactive metal become used up in time. If it is not replaced regularly the iron will eventually start to rust.

In a similar way iron and steel can be protected from rusting by attaching them to the **negative terminal** of a **power supply**. This works because the negative terminal pushes electrons into the iron, which again makes it more difficult for the iron to lose electrons itself. This method of protection has been used on oil-drilling platforms in the North Sea.

Looking at different surface barriers

All surface barriers work in the same way. However, they each have advantages and disadvantages.

Paint, grease and plastic

These coatings are fairly cheap and easy to apply but they can easily be broken and rusting will occur as water and oxygen get at the metal.

Tin-plating

Tin is unreactive, strong and non-poisonous so it is useful for coating food 'tins'. However if the coating becomes scratched or broken the iron underneath will rust faster than normal, as electrons flow from the iron to the less reactive tin.

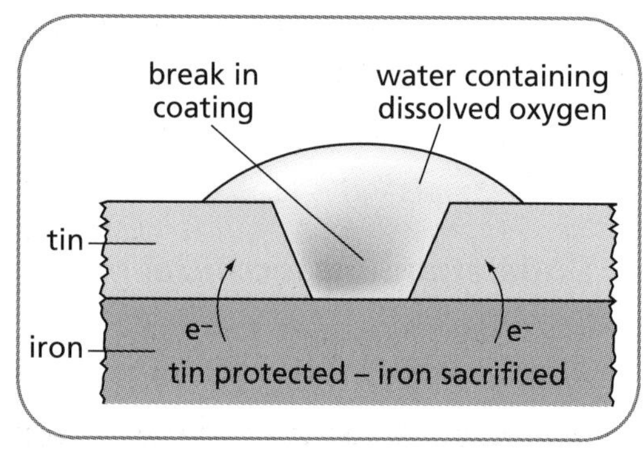

Figure 12.5 Contact with tin speeds up the rusting of iron

Galvanising

Coating iron with zinc is called **galvanising** and it protects in two ways. First it acts as a surface barrier to keep out air and water. Then, if the coating becomes scratched or broken, it will protect sacrificially by giving electrons to the less reactive iron.

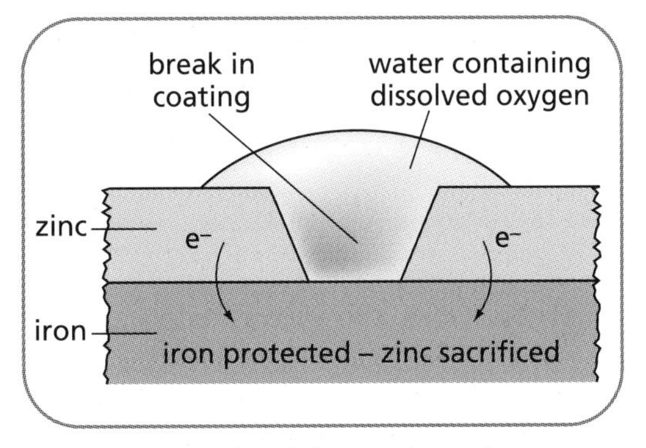

Figure 12.6 Galvanising protects in two ways

Electroplating

Electroplating involves putting a thin coating of an expensive metal like chromium, nickel or silver onto iron by using electrolysis. This produces a fairly cheap, strong, unreactive barrier to keep out air and water. However, if the coating becomes scratched, like tin-plating, the iron rusts faster than normal as electrons flow away from the iron.

Summary

- Corrosion is a reaction that forms a compound at the surface of a metal.
- During corrosion the metal is oxidised, $M \rightarrow M^+ + e^-$.
- The more reactive the metal the faster it corrodes.
- Rusting is the corrosion of iron.
- Water, oxygen and ions must be present for rusting to occur.
- The presence of an electrolyte, like salt, increases the rate of rusting.
- Ferroxyl indicator changes green to blue with Fe^{2+} and green to pink with OH^-.
- Rusting is a redox reaction and can be used to set up a cell.
- During rusting the iron is oxidised: $Fe \rightarrow Fe^{2+} + 2e^-$ then $Fe^{2+} \rightarrow Fe^{3+} + e^-$.
- The water and oxygen are reduced during rusting: $2H_2O + O_2 + 4e^- \rightarrow 4OH^-$.
- A surface barrier prevents rusting by keeping out air and water.
- Examples of surface barriers include paint, plastic, grease, tin, zinc and electroplating.
- Sacrificial protection works by attaching iron to a metal higher up the electrochemical series, e.g. magnesium.
- The metal higher up the electrochemical series corrodes first and gives electrons to the iron.
- Galvanising, coating in zinc, works in two ways, as a surface barrier and by sacrificial protection.
- Attaching iron to the negative terminal of a power supply also protects iron.
- When less reactive metals are in contact with iron, the iron rusts faster.

Chapter 13

PLASTICS AND SYNTHETIC FIBRES

Everywhere you look plastics and synthetic fibres are being used in place of natural materials. But what are plastics, how are they made and why have they become so widely used? This chapter deals with the structure and properties of plastics and synthetic fibres. It introduces the process of polymerisation and briefly looks at the impact of plastics on our environment.

Key Words

★ **addition polymer** ★ **biodegradable** ★ **monomer** ★ **plastic** ★ **polymer** ★ **polymerisation** ★ **repeating unit** ★ **synthetic fibre** ★ **thermoplastic** ★ **thermosetting**

Properties of plastics

Most **plastics** and **synthetic fibres** are examples of petrochemicals, that is, they are made from molecules obtained from **crude oil**.

Note that 'synthetic' fibres refer to fibres that are man-made. They have not been obtained directly from plants or animals.

Chemists have developed a range of plastics and synthetic fibres to replace natural materials like wood, metal, wool and cotton. These new materials are not only cheaper but they also have special properties that make them suitable for a variety of uses.

Plastics and **synthetic fibres** are generally:

- Flexible, light and strong.
- Durable and resist corrosion.
- Good insulators of heat.
- Good insulators of electricity.

For example, the plastic coating for electrical cables is a durable, flexible electrical insulator. Synthetic fibres used for clothing are strong, light and help keep in heat.

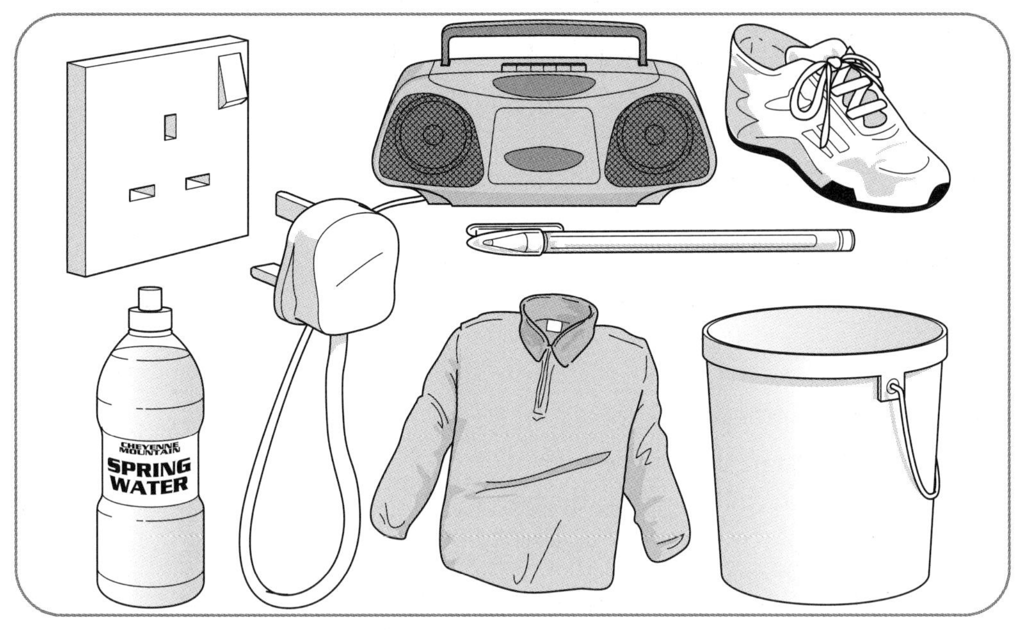

Figure 13.1 Plastics and synthetic fibres

Disadvantages of using plastics

Most plastics and synthetic fibres are not **biodegradable** which means they do not break down in nature. Being durable means they can cause long term pollution problems.

Most plastics and synthetic fibres are **flammable** and give off **poisonous** fumes when they burn. Some examples of toxic gases produced by burning plastics are shown below.

Gases produced by burning plastics

Toxic gas	Where formed
carbon monoxide	incomplete combustion of any plastic
hydrogen chloride	burning polychloroethene, which contains hydrogen and chlorine
hydrogen cyanide	burning polyurethane foams, which contain hydrogen carbon and nitrogen

Most deaths in house fires are caused by breathing in smoke and toxic fumes, rather than the heat and flames.

Polymers

All plastics and synthetic fibres are examples of **polymers**. These are substances that contain large molecules made by joining thousands of small molecules together. The small molecules are called **monomers** and the reaction that joins them together is **polymerisation**.

Addition polymers

Addition polymers are formed from monomers which contain a C=C double bond. These monomers are often obtained by the catalytic cracking of hydrocarbon fractions from crude oil.

One of the first addition polymers made was polyethene, commonly called polythene. The polymerisation of ethene molecules is shown below.

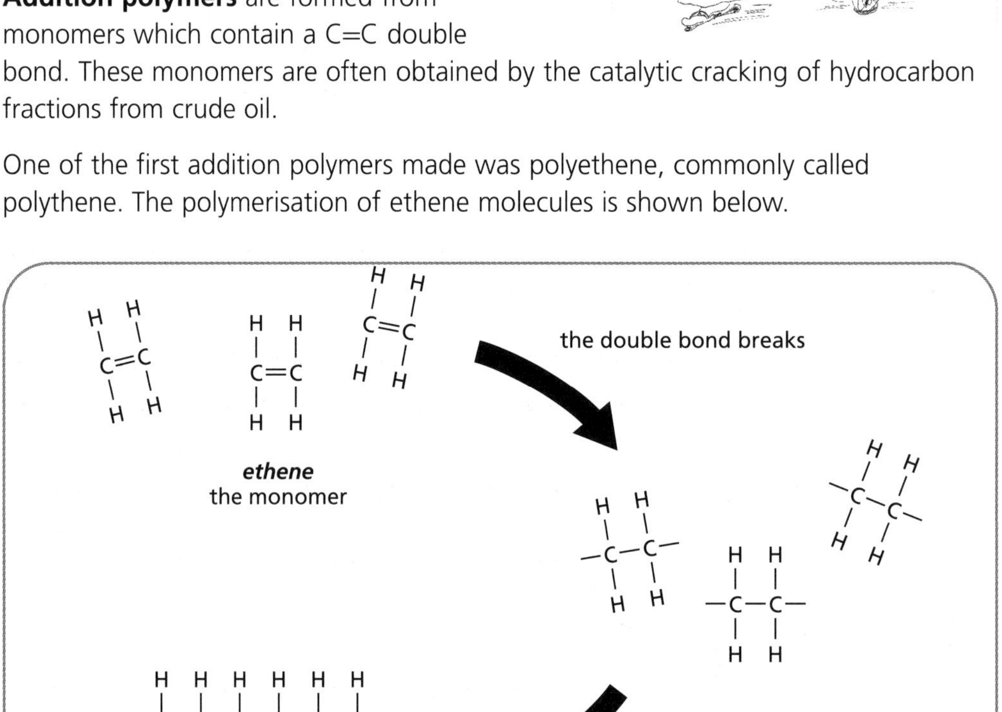

Figure 13.2 Forming polyethene

There are many other examples of addition polymers. Given suitable information, you should be able to draw the structures of the polymer chain, monomer, or **repeating unit** of an addition polymer. Examples of these structures are shown in the table on the next page.

Some examples of polymers

Polymer	Monomer	Polymer chain	Repeating unit
polychloroethene	$\begin{array}{c}H\\|\\C\\|\\H\end{array}{=}\begin{array}{c}H\\|\\C\\|\\Cl\end{array}$ chloroethene	$-\begin{array}{c}H\\|\\C\\|\\H\end{array}-\begin{array}{c}H\\|\\C\\|\\Cl\end{array}-\begin{array}{c}H\\|\\C\\|\\H\end{array}-\begin{array}{c}H\\|\\C\\|\\Cl\end{array}-\begin{array}{c}H\\|\\C\\|\\H\end{array}-\begin{array}{c}H\\|\\C\\|\\Cl\end{array}-$	$-\begin{array}{c}H\\|\\C\\|\\H\end{array}-\begin{array}{c}H\\|\\C\\|\\Cl\end{array}-$
polytetrafluoroethene	$\begin{array}{c}F\\|\\C\\|\\F\end{array}{=}\begin{array}{c}F\\|\\C\\|\\F\end{array}$ tetrafluoroethene	$-\begin{array}{c}F\\|\\C\\|\\F\end{array}-\begin{array}{c}F\\|\\C\\|\\F\end{array}-\begin{array}{c}F\\|\\C\\|\\F\end{array}-\begin{array}{c}F\\|\\C\\|\\F\end{array}-\begin{array}{c}F\\|\\C\\|\\F\end{array}-\begin{array}{c}F\\|\\C\\|\\F\end{array}-$	$-\begin{array}{c}F\\|\\C\\|\\F\end{array}-\begin{array}{c}F\\|\\C\\|\\F\end{array}-$
polystyrene	$\begin{array}{c}H\\|\\C\\|\\H\end{array}{=}\begin{array}{c}H\\|\\C\\|\\C_6H_5\end{array}$ styrene	$-\begin{array}{c}H\\|\\C\\|\\H\end{array}-\begin{array}{c}H\\|\\C\\|\\C_6H_5\end{array}-\begin{array}{c}H\\|\\C\\|\\H\end{array}-\begin{array}{c}H\\|\\C\\|\\C_6H_5\end{array}-\begin{array}{c}H\\|\\C\\|\\H\end{array}-\begin{array}{c}H\\|\\C\\|\\C_6H_5\end{array}-$	$-\begin{array}{c}H\\|\\C\\|\\H\end{array}-\begin{array}{c}H\\|\\C\\|\\C_6H_5\end{array}-$

Hints and Tips

1. Adding 'poly' in front of the name of the monomer gives the polymers' name.
2. The repeating unit looks like the monomer, with a single C–C bond in place of the C=C double bond.
3. The structural formulas of the polymer and repeating unit show the bonds continue at both ends.

Types of polymer

Plastics and synthetic fibres can be put into two groups, depending on how they are affected by heat.

Thermoplastics, like polythene and polychloroethene, soften and melt on heating. These plastics are easily moulded into different shapes. Thermoplastic polymers consist of separate long-chain molecules that can be easily separated from each other.

Figure 13.3 Molecules of a thermoplastic plastic

Thermosetting plastics, like bakelite, do *not* soften or melt on heating. These plastics set hard when formed. Thermosetting polymers consist of long-chain molecules that are held in a rigid structure by cross-links between the chains.

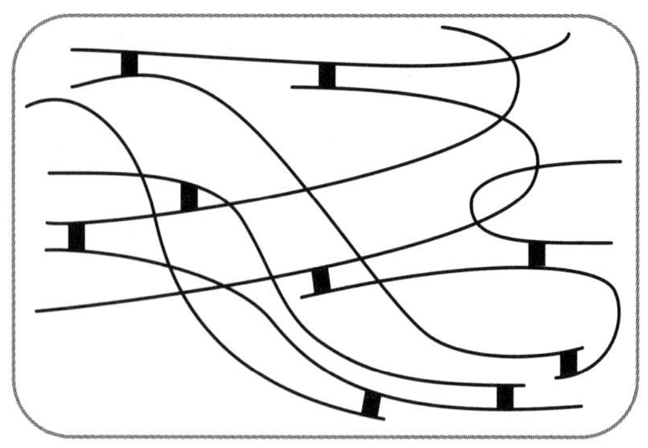

Figure 13.4 Molecules of a thermosetting plastic

Summary

- Crude oil is the main source of most plastics and synthetic (man-made) fibres.
- The main advantages of using plastics and synthetic fibres are that they are strong, light, durable, flexible and good insulators of heat and electricity.
- The main disadvantages of using plastics and synthetic fibres are that they are non-biodegradable and they burn forming poisonous fumes.
- Polymers are formed by joining lots of small molecules, called monomers, together.
- The monomers for addition polymers all contain a C=C double bond.
- The simplest addition polymer, polyethene is formed from ethene.
- During addition polymerisation the double bond breaks and the monomers join together.
- The repeating unit in an addition polymer looks like the monomer without the C=C double bond.
- Thermoplastic polymers can melt on heating and be reshaped.
- Thermosetting polymers do not melt on heating due to cross links between the chains

Chapter 14

FERTILISERS AND NITROGEN CHEMISTRY

As the world's population increases we need to find ways of growing crops more efficiently. But what do plants need to grow, and what can the chemical industry do to help us grow more food? This chapter looks at the nutrients required for plant growth and the manufacture and use of synthetic fertilisers.

Key Words

★ **fertiliser** ★ **fixed nitrogen** ★ **free nitrogen** ★ **Haber process**
★ **nitrogen-fixing (nitrifying) bacteria** ★ **nitrogen cycle**
★ **Ostwald process** ★ **synthetic fertiliser** ★ **plant nutrient**
★ **reversible reaction**

Plant nutrients

Plants need certain elements in the soil for healthy growth. The three most important of these **plant nutrients** are **nitrogen**, **phosphorus** and **potassium**. These elements must be present as soluble compounds, so they can be absorbed through the plants' roots. In nature these elements are continually being removed and replaced in a **nutrient cycle** that keeps the soil fertile and able to support plant growth.

This nutrient cycle is often broken, and crops and animals are removed to feed our growing population. This means that the soil becomes less fertile and **fertilisers** are needed to replace the lost nutrients. Fertilisers contain the elements essential for plant growth.

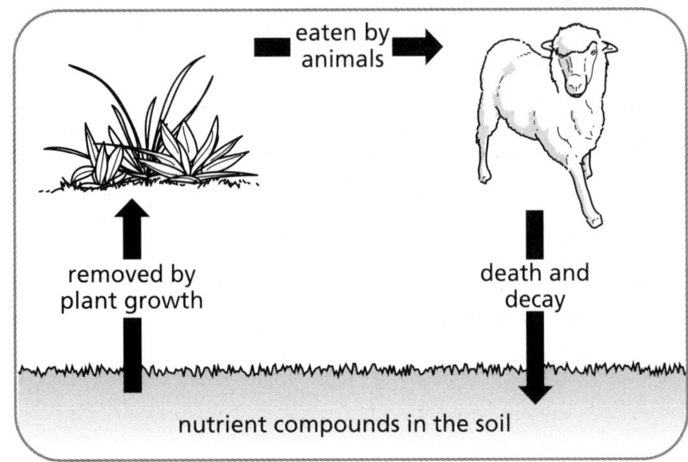

Figure 14.1 The nutrient cycle

There are two main sources. Natural fertilisers like **compost** and **manure**, decomposing plant and animal material, are excellent sources of plant nutrients. There is, however, a limited supply of natural fertilisers. So **synthetic fertilisers**, man-made compounds that contain the essential elements, are often needed to replace nutrients lost from the soil.

Synthetic fertilisers

Some examples of compounds used in synthetic fertilisers are shown. They are suitable for this use as they are soluble and contain the elements needed for plant growth.

Example

Some synthetic fertilisers:

Potassium nitrate – $K^+ NO_3^-$

Ammonium sulphate – $(NH_4^+)_2 SO_4^{2-}$

Calcium phosphate – $(Ca^{2+})_3 (PO_4^{3-})_2$

Ammonium nitrate – $NH_4^+ NO_3^-$

As different plants require different amounts of each nutrient, a range of fertilisers is available. They are often called **'NPK' fertilisers** as they contain the main plant nutrients, nitrogen (N), phosphorus (P) and potassium (K).

Figure 14.2 Using an NPK fertiliser

Pollution problems

As synthetic fertilisers are water soluble, their overuse can lead to problems of pollution in rivers and lochs.

- High levels of phosphate in lochs cause increased algae growth, called 'algal blooms'. These algae can damage and kill many forms of water life.
- An increased level of nitrate in drinking water is thought to cause serious health problems.

The use of fertilisers in farming is therefore closely monitored and controlled.

Nitrogen and living things

All living things need nitrogen. Plants need nitrogen to make compounds called amino acids, which are used to make **proteins**. Animals get their nitrogen by eating plants or other animals. The atmosphere contains nearly 80% nitrogen gas but most plants cannot use this **'free' nitrogen**. Most plants need **'fixed' nitrogen**, compounds like ammonium nitrate, in the soil. The way nitrogen passes from one living thing to another and from free nitrogen in the air to fixed nitrogen in the soil is called the **nitrogen cycle**. The nitrogen cycle is a very important life cycle, which contains many complex changes.

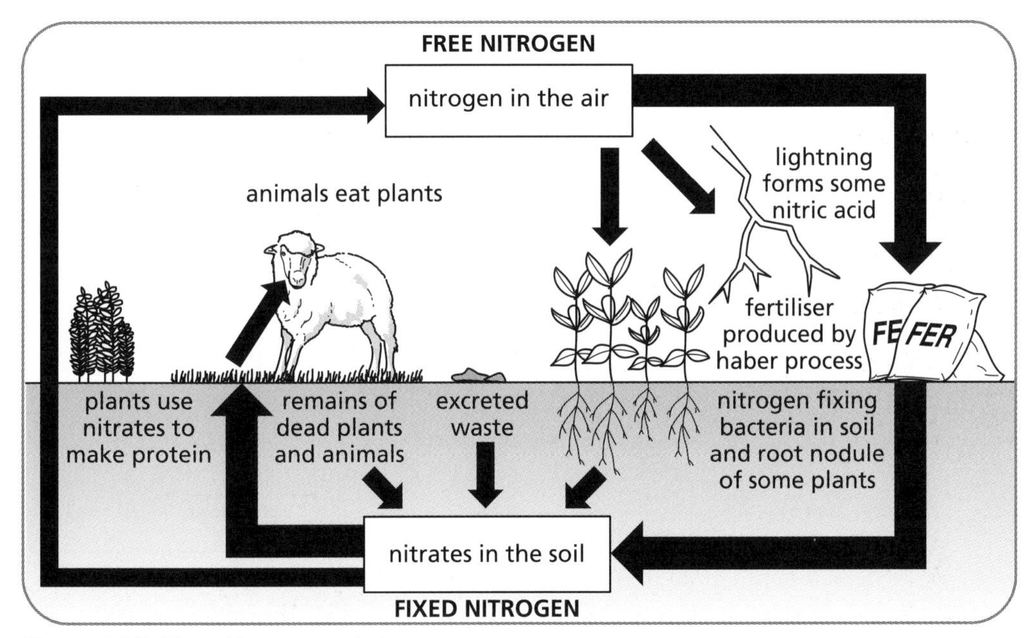

Figure 14.3 The nitrogen cycle

To keep the nitrogen cycle in balance farmers often have to use expensive synthetic fertilisers made by the chemical industry. For example, ammonium nitrate and ammonium sulphate, which are used as fertilisers, are made from ammonia manufactured by the **Haber process**.

A cheaper way of fixing nitrogen is to grow **leguminous plants** like peas, beans and clover. These plants have special **nitrogen-fixing (nitrifying) bacteria** in root nodules. They can be used to return nitrogen compounds to the soil, but a year's crop production is lost in the process.

Nitrogen and its compounds

Almost 80% of the atmosphere is made up of nitrogen. This unreactive gas has diatomic molecules held together by strong triple bonds.

Nitrogen dioxide and nitric acid

The nitrogen molecule is very stable, so nitrogen does not react or form compounds easily. To make nitrogen react we need to use high voltage electricity. In 'sparking air' the electrical energy is needed to break up the nitrogen molecules, allowing them to react with the oxygen to form **nitrogen dioxide**.

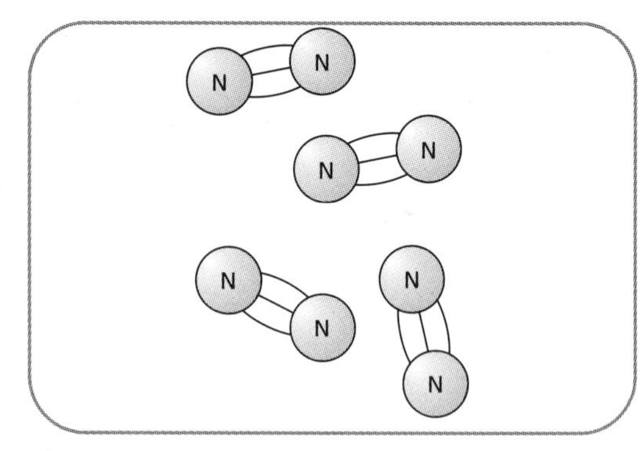

Figure 14.4 Nitrogen gas

$$N_2(g) + 2O_2(g) \rightarrow 2NO_2(g)$$

nitrogen + oxygen → nitrogen dioxide

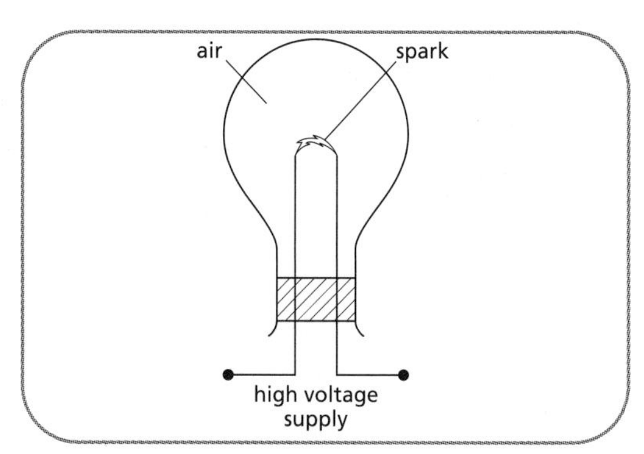

Figure 14.5 Sparking air

The nitrogen dioxide can be detected, as it is a **brown gas** that dissolves in water to produce an acidic solution of nitric acid.

$$4NO_2(g) + 2H_2O(l) + O_2(g) \rightarrow 4HNO_3(aq)$$

nitrogen dioxide + water + oxygen → nitric acid

The same reactions occur during lightning storms and in car engines, where the spark plug supplies the energy needed to start the reaction. In both cases nitrogen dioxide is produced, and this can dissolve in rainwater to form nitric acid.

This means that car engines and lightning storms help to return some fixed nitrogen to the soil. However, they also unfortunately contribute to acid rain by producing nitric acid.

Note that although high voltage electricity can be used to manufacture nitric acid it is too expensive to be an economic industrial process.

Ammonia

Ammonia formula, NH_3, is one of the most important nitrogen compounds.

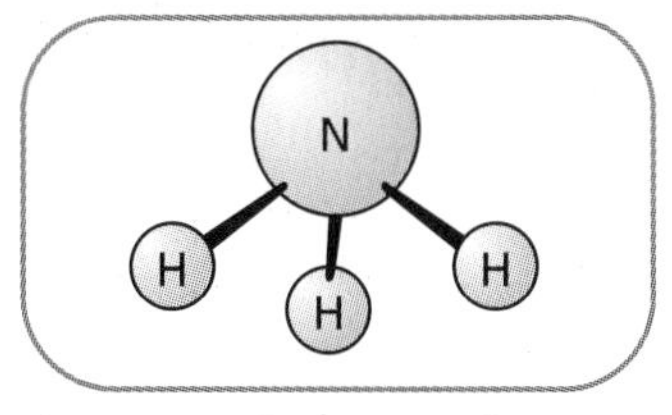

Figure 14.6 Ammonia – NH_3

Ammonia can be made in the laboratory by heating an ammonium salt with an alkali. For example, if ammonium chloride is heated with sodium hydroxide the following reaction occurs.

$$NH_4Cl(s) + NaOH(s) \rightarrow NH_3(g) + NaCl(s) + H_2O(l)$$

ammonium chloride + sodium hydroxide → ammonia + sodium chloride + water

The main properties of ammonia are:

- colourless gas
- pungent smell
- very soluble in water
- forms alkaline solutions.

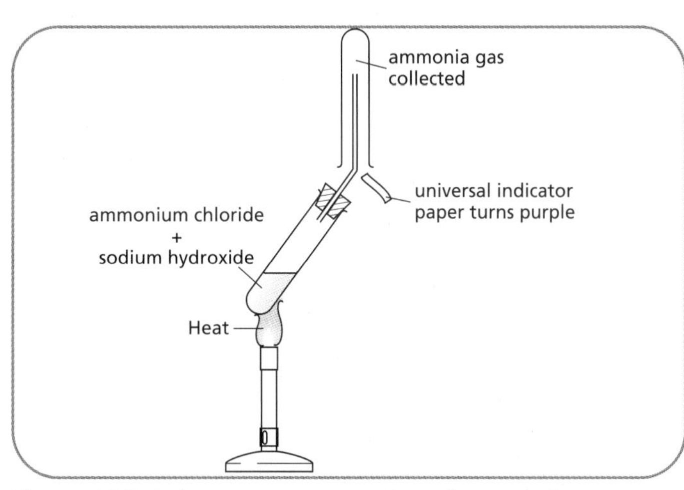

Figure 14.7 Laboratory preparation of ammonia

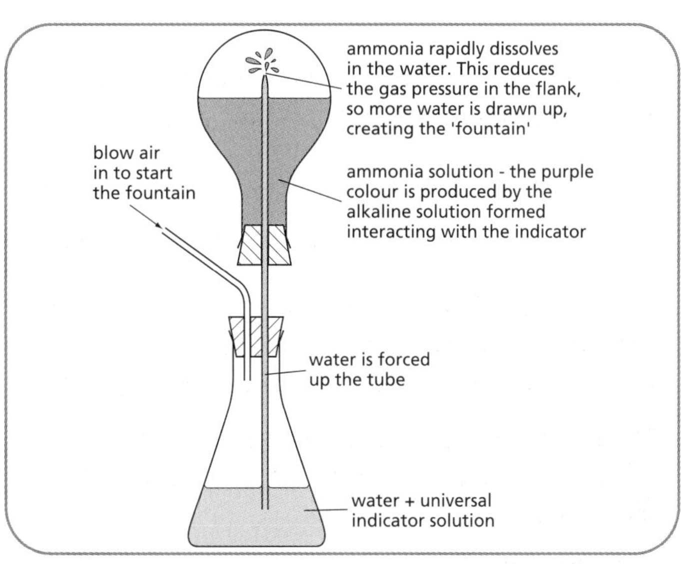

Figure 14.8 The fountain experiment shows that ammonia is a very soluble gas

The Haber process

The Haber process is used to manufacture thousands of tonnes of ammonia each year. The main uses of this ammonia are to make fertilisers and nitric acid.

The raw materials for the Haber process are fairly cheap and easy to obtain. Nitrogen is obtained from the air and hydrogen is obtained from steam and methane. The overall reaction equation for the process is shown on the next page.

This symbol, $\rightleftharpoons$, tells you that the reaction forming ammonia is a **reversible reaction**. This means that it is impossible to get 100% yield of products. So the conditions have to be controlled carefully to get the most economic yield of ammonia possible.

Equation To Learn

$$N_2(g) + 3H_2(g) \rightleftharpoons 2NH_3(g)$$

nitrogen + hydrogen $\rightleftharpoons$ ammonia

The conditions chosen for the Haber process are:

- temperature 400 °C
- pressure 200 atmospheres
- catalyst of iron granules.

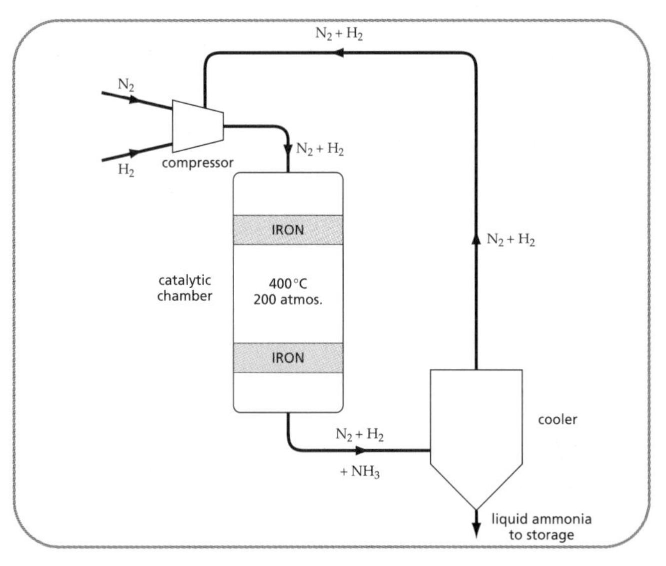

Figure 14.9 The Haber process

Explanation of the conditions chosen for the Haber process

- The yield of ammonia is greatest at low temperature. However, if the temperature is too low then the reaction is too slow. Therefore a moderate temperature of **400 °C** is used.
- The yield of ammonia is greatest at high pressures. However, using very high pressures is expensive. Therefore a pressure of **200 atmospheres** is used.

- A **catalyst of iron** is used to speed up the reaction. It is broken up into small pieces to increase the surface area.

These conditions produce a yield of about 15% ammonia. However, none of the reactants are wasted. Cooling turns the ammonia into a liquid, so it can be removed and the unreacted nitrogen and hydrogen are recycled back into the converter.

The Oswald process

Most of the nitric acid made in Britain is used to make fertilisers and explosives. In this country the **Ostwald process** is used to make nitric acid.

This process starts with the **catalytic oxidation of ammonia** to form first **nitrogen monoxide**, then **nitrogen dioxide**.

$$4NH_3(g) + 5O_2(g) \rightarrow 4NO(g) + 6H_2O(l)$$

ammonia + oxygen → nitrogen monoxide + water

$$2NO(g) + O_2(g) \rightarrow 2NO_2(g)$$

nitrogen monoxide + oxygen → nitrogen dioxide

The nitrogen dioxide is then dissolved in water, with more oxygen, to form nitric acid.

$$4NO_2(g) + 2H_2O(l) + O_2(g) \rightarrow 4HNO_3(aq)$$

nitrogen dioxide + water + oxygen → nitric acid

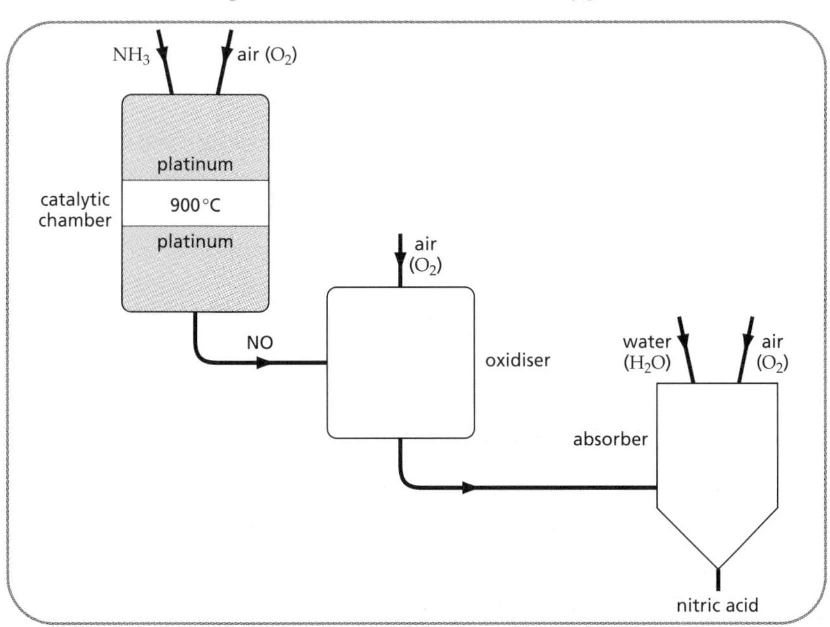

Figure 14.10 The Ostwald process

The conditions chosen for the Ostwald process are:

- temperature of 900 °C
- catalyst of platinum.

A moderate temperature of 900 °C is needed for the process. However, as the reaction is exothermic there is no need to keep heating after it has started.

The **platinum catalyst** speeds up the reaction. The catalyst is in the form of fine gauze to increase the surface area.

The catalytic oxidation of ammonia can be carried out in the laboratory as shown opposite. The hot platinum catalyst starts the reaction. However, no further heating is required and the catalyst will continue to glow as the oxidation reaction is exothermic.

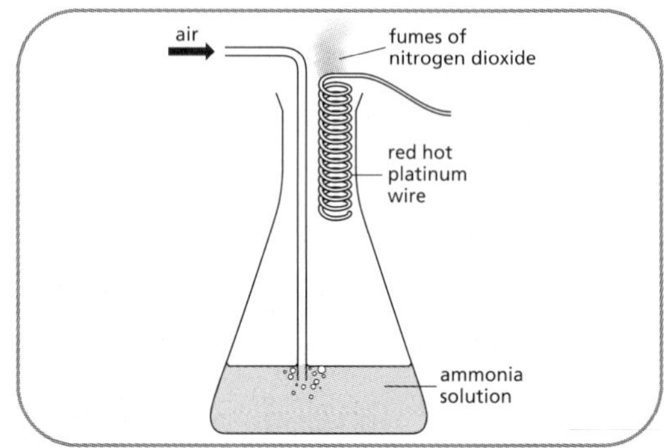

Figure 14.11 The laboratory catalytic oxidation of ammonia

Reactions of ammonia and nitric acid

Neutralisation reactions are used to produce **salts** of ammonia and nitric acid. Ammonia forms ammonium salts when it reacts with acids. Nitric acid forms nitrate salts when it reacts with bases. For example, if ammonia is bubbled through nitric acid neutralisation occurs.

$$NH_3(g) + HNO_3(aq) \rightarrow NH_4NO_3(aq)$$

ammonia + nitric acid → ammonium nitrate

Other examples of salts which could be produced by neutralisation reactions involving ammonia or nitric acid are: $(NH_4)_2SO_4$, NH_4NO_3 and $(NH_4)_3PO_4$. All of these salts can be used as fertilisers.

Summary

- The three main plant nutrients are nitrogen, phosphorus and potassium.
- Synthetic fertiliser are soluble compounds that contain N, P and K.
- We need to make more fertilisers as the world's population is increasing.
- Using fertiliser can cause algal blooms on lochs and poison water supplies.
- The nitrogen cycle describes the changes between 'free' nitrogen in air and 'fixed' nitrogen in compounds in the soil.
- Leguminous plants, like clover, have bacteria in root nodules which 'fix' nitrogen.
- Industrially nitrogen is fixed by the Haber process, which makes ammonia.
- Using bacterial methods for fixing nitrogen is cheaper than industrial methods.
- Ammonia: colourless gas, smelly, very soluble, alkaline, used to make fertilisers.
- The Haber process: $N_2 + 3H_2 \rightleftharpoons 2NH_3$ (N_2 from air and H_2 from H_2O and CH_4).
- The conditions for the Haber process are: 400 °C, 200 atmospheres, iron catalyst and recycle unreacted nitrogen and hydrogen.
- Ammonia can be made in the laboratory by heating an ammonium salt with a solid alkali.
- Nitric acid is made by the catalytic oxidation of ammonia.
- The Ostwald process is: $NH_3 + O_2 \rightarrow NO + H_2O$ at 900 °C and a platinum catalyst, then: $NO + O_2 \rightarrow NO_2$ followed by: $NO_2 + H_2O + O_2 \rightarrow HNO_3$.
- Nitrogen dioxide: brown gas, soluble in water, forms nitric acid, used to make fertilisers.
- During sparking air and lightning: $N_2 + 2O_2 \rightarrow 2NO_2$; too expensive for manufacture.
- Neutralisation of ammonia forms ammonium salts which can be used as fertilisers.
- Neutralisation of nitric acid forms nitrate salts which can be used as fertilisers.

Chapter 15

CARBOHYDRATES AND RELATED SUBSTANCES

Biochemistry is the study of the molecules and reactions which occur in living things, including the food you eat. But what kind of chemicals make up our food and what kinds of changes are involved in their manufacture and use? This chapter deals with the chemistry of carbohydrates, an important food class, looking at their molecular structure, key reactions and some related compounds.

Key Words

★ **alcohol** ★ **alkanol** ★ **carbon cycle** ★ **chlorophyll**
★ **condensation reaction** ★ **condensation polymer** ★ **digestion**
★ **disaccharide** ★ **enzyme** ★ **fermentation** ★ **global warming**
★ **greenhouse effect** ★ **hydrolysis reaction** ★ **monosaccharide**
★ **photosynthesis** ★ **polysaccharide** ★ **respiration**

Introduction to carbohydrates

Foods like pasta, bread and potatoes are all rich in **carbohydrates**. These are naturally occurring compounds that are made up of **carbon**, **hydrogen** and **oxygen** only. Carbohydrates are often described as energy foods. We need carbohydrates in our diet as they provide the energy for movement, warmth, growth and repair.

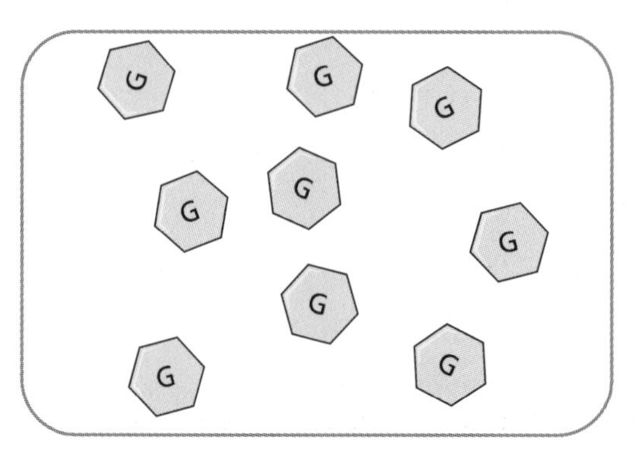

Figure 15.1 Glucose molecules

There are three main classes of carbohydrate.

Common carbohydrate	Molecular formula
glucose	$C_6H_{12}O_6$
fructose	$C_6H_{12}O_6$
sucrose	$C_{12}H_{22}O_{11}$
maltose	$C_{12}H_{22}O_{11}$
starch	$(C_6H_{10}O_5)_n$
cellulose	$(C_6H_{10}O_5)_n$

- **Monosaccharides**, simple sugars like **glucose** and **fructose**. These are isomers as they have the same molecular formula, $C_6H_{12}O_6$, but different structural formulas.
- **Disaccharides**, like **sucrose** and **maltose**. These are also isomers, with the same molecular formula $C_{12}H_{22}O_{11}$. Disaccharides are made from two monosaccharide molecules joined together.
- **Polysaccharides**, like **starch** and **cellulose**, which have the molecular formula $(C_6H_{10}O_5)_n$. These are natural polymers made by joining many monosaccharide molecules together.

Glucose and sucrose, which are often found in 'sweet' foods, are soluble in water. Starch, which is commonly found in vegetables like potatoes, is not sweet or soluble. Cellulose, also insoluble in water, is the structural material of plants and is the main constituent of cotton.

Testing carbohydrates

There are two main tests used for carbohydrate foods.

- **Iodine solution** turns from brown to blue/black in the presence of starch.
- **Benedict's** or **Fehling's solution** changes colour from blue to orange when heated with glucose, fructose or maltose.

Note that neither iodine nor Benedict's solution changes colour with sucrose.

Photosynthesis

Green plants manufacture carbohydrates, like glucose, from carbon dioxide and water by a process called **photosynthesis**.

For photosynthesis to take place the following must be present.

- **Carbon dioxide** and **water**, the raw materials or reactants in the process.
- **Sunlight**, the source of energy, which is needed for the reaction to occur.
- **Chlorophyll**, substance that gives plants their green colour, which traps the sun's energy.

During photosynthesis the carbon dioxide and water are converted into glucose and oxygen. The glucose stores the sun's energy and the oxygen is released into the atmosphere.

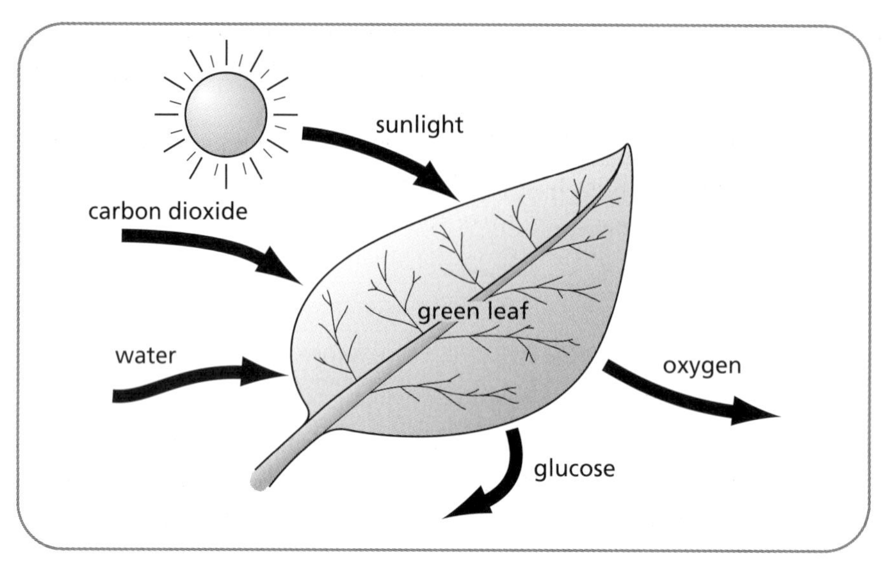

Figure 15.2 Photosynthesis

Equations To Learn

$$6CO_2(g) + 6H_2O(l)\ (+\text{energy}) \rightarrow C_6H_{12}O_6(s) + 6O_2(g)$$

carbon dioxide + water → glucose + oxygen

Respiration

Plants and animals obtain the energy they needed for life from the energy stored in carbohydrates. The process which releases the energy is called **respiration**. It is the opposite of photosynthesis, as it combines glucose and oxygen to produce carbon dioxide, water and energy.

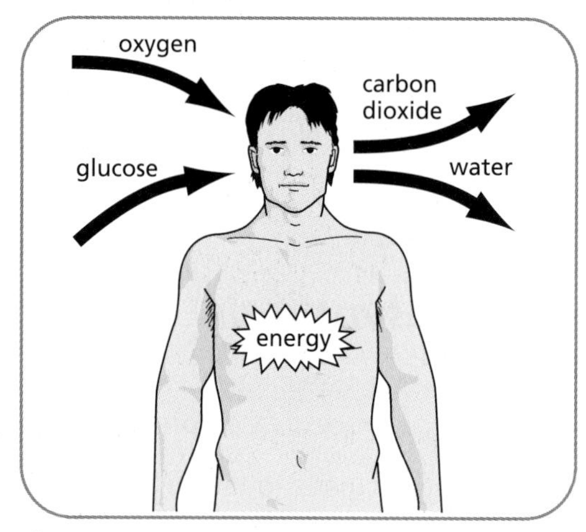

Figure 15.3 Respiration

Equation To Learn

$$C_6H_{12}O_6(s) + 6O_2(g) \rightarrow 6CO_2(g) + 6H_2O(l)\ (+\text{energy})$$

glucose + oxygen → carbon dioxide + water

The same overall reaction occurs during the combustion of carbohydrates. However, in combustion the energy release is quicker and less controlled than in respiration.

The production of carbon dioxide and water on combustion is *not* a test for carbohydrates. Carbohydrates, hydrocarbons or any compound that contains carbon and hydrogen will form carbon dioxide and water when they burn.

The carbon cycle

Photosynthesis and respiration are key reactions in the important life cycle called the **carbon cycle**. This cycle describes how carbon changes between carbon dioxide in the atmosphere and carbon compounds in living things and fossil fuels.

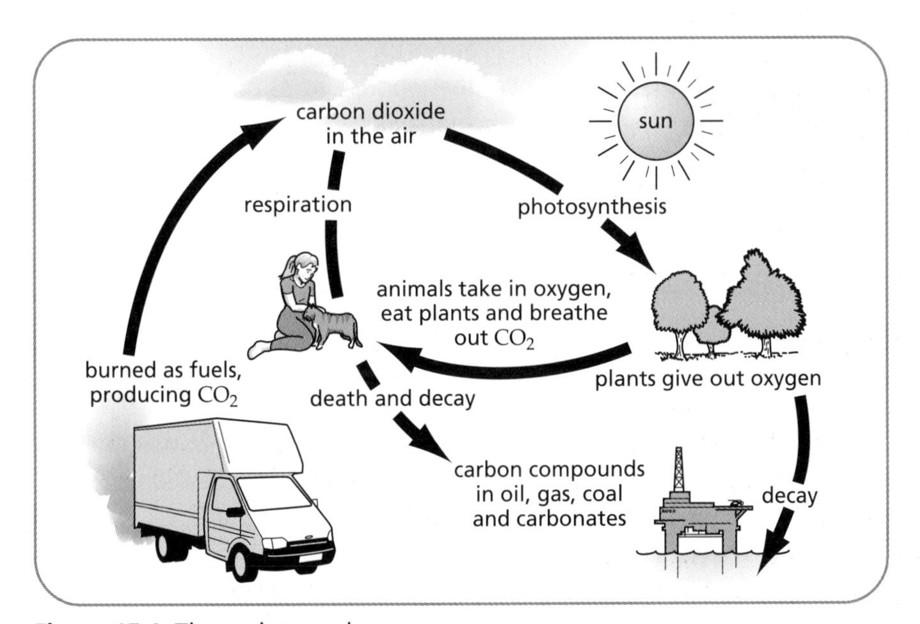

Figure 15.4 The carbon cycle

In the past the amount of carbon dioxide removed from the atmosphere has been balanced by the amount returned. However, this balance is now in danger. Human activity is causing the concentration of carbon dioxide in the atmosphere to rise. Firstly we are burning more fossil fuels, which produces carbon dioxide. We are also destroying large areas of forest that use up the carbon dioxide during photosynthesis.

Carbon dioxide, and other gases, help trap the sun's energy by what's called the **greenhouse effect**. With carbon dioxide levels rising, some scientists believe that **global warming** will occur, and that the average temperatures around the earth will increase.

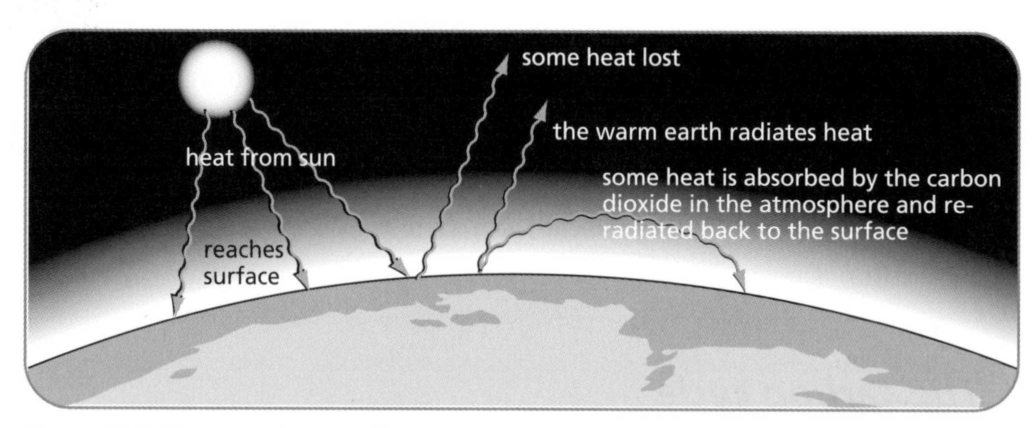

Figure 15.5 The greenhouse effect

Average world temperatures appear to be increasing already and some scientists predict an average rise of up to 5 °C in the next hundred years. If this type of global warming occurs, it would have a enormous affect on our environment. Weather patterns would change dramatically all over the world, and as the polar ice caps melt, sea levels would rise, flooding many costal areas.

Carbohydrate classes

The main difference between the carbohydrate classes – monosaccharides, disaccharides and polysaccharides – is the size of their molecules. This can be shown by a simple experiment. If a beam of light is shone through solutions of glucose, sucrose and starch, the beam can only be seen passing through the starch solution. The starch molecules are much larger than the molecules of glucose and sucrose. As a result the starch molecules reflect the light and can be seen.

Building up carbohydrates

Photosynthesis forms simple sugars like glucose. The larger carbohydrates have to be made by joining these simple sugars together. For example, if two glucose molecules are joined together a maltose molecule is formed.

This is called a **condensation reaction**, as a water molecule is removed to join the two glucose molecules together.

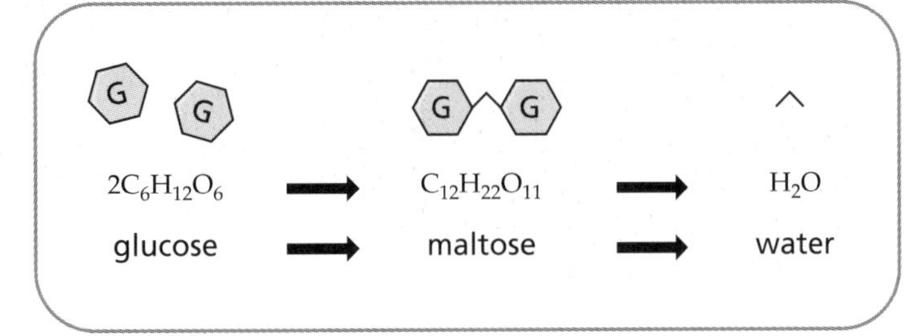

Figure 15.6 Forming maltose by condensation

Starch is also formed by a condensation reaction in which thousands of glucose molecules are joined together to make the polysaccharide.

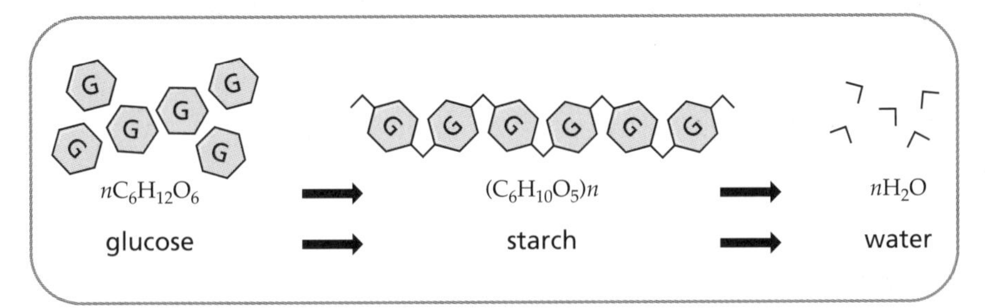

Figure 15.7 Making starch by condensation

This is not only a **condensation reaction** it is also a **polymerisation reaction**, as lots of small molecules have been joined together to make a long-chain molecule. Starch is therefore a natural **condensation polymer**, formed from the monomer glucose by **condensation polymerisation**.

Breaking up carbohydrates

During **digestion**, starch molecules are broken down into glucose, so the smaller molecules can pass through the gut wall and get into the blood. The starch can be broken down by dilute acids, like hydrochloric acid, or amylase, a substance found in saliva.

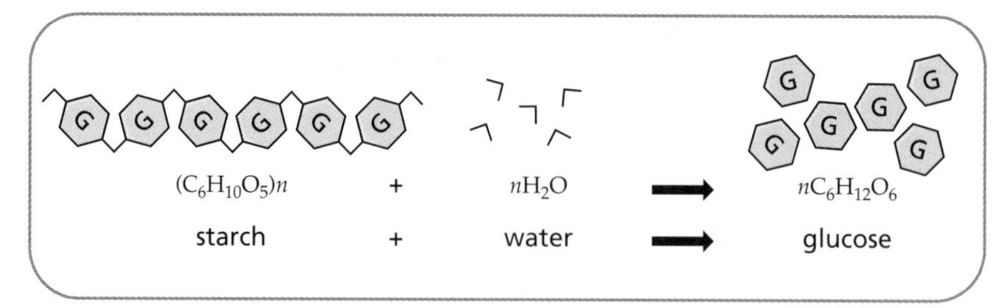

Figure 15.8 The hydrolysis of starch

This is called a **hydrolysis reaction**. It is the opposite of a condensation as water is added to split up large molecules.

These reactions can be brought about in the laboratory as well as the body. Amylase, which is a biological catalyst or **enzyme**, breaks down starch more quickly than hydrochloric acid. When amylase is used the reaction is carried out at 37 °C, body temperature.

Alcohol and fermentation

For thousands of years we have made alcoholic drinks from different fruits and vegetables. The proper name for the **alcohol** in these drinks is **ethanol**. This is a compound of carbon, hydrogen and oxygen, which has a similar structure to ethane.

Alcoholic drinks are made by mixing yeast and water with a crushed fruit or vegetable. The process is called **fermentation** and involves breaking down glucose into ethanol and carbon dioxide.

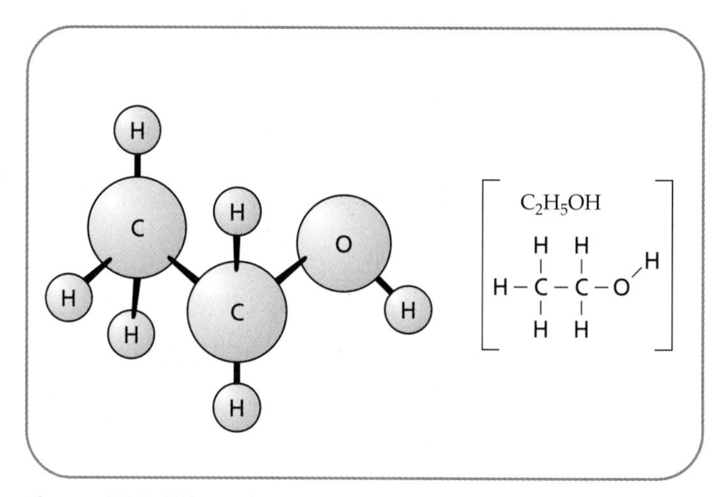

Figure 15.9 Ethanol

The **enzyme** from yeast, **zymase**, acts as a catalyst for the reaction. So the temperature and pH must be carefully controlled to help the enzyme work at its best.

Equation To Learn

$$C_6H_{12}O_6 \rightarrow 2C_2H_5OH + 2CO_2$$

glucose → ethanol + carbon dioxide

Fermentation can only produce solutions of up to 12% ethanol. Above this concentration the yeast stops working. More concentrated alcoholic drinks, called spirits, are produced by **distillation**. This works as the ethanol and water have different boiling points. During distillation the ethanol evaporates first so its concentration in the distillate is higher than in the original solution.

Boiling point of ethanol = 80 °C
Boiling point of water = 100 °C

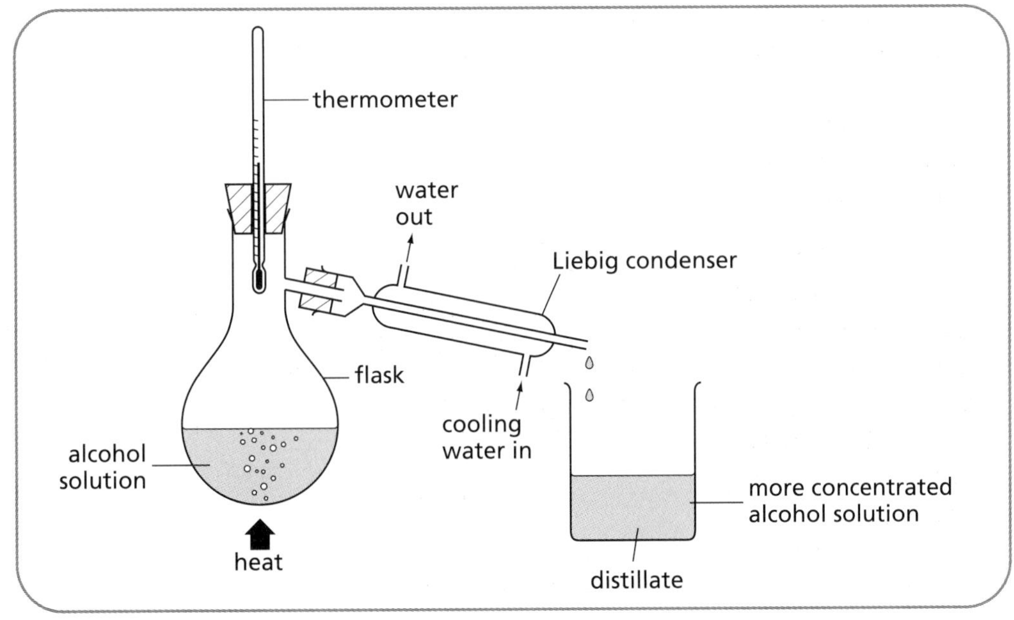

Figure 15.10 Distillation

The type of alcoholic drink produced depends on the % alcohol and the plant source of the carbohydrate.

Some alcoholic drinks

Alcoholic drink	Plant source	% Alcohol	Distilled?
whisky	barley	40%	yes
wine	grapes	5–10%	no
beer	barley	3–6%	no
vodka	potatoes	35%	yes

Enzymes

Enzymes are sometimes described as biological catalysts. They speed up reactions in living things. Each enzyme can only do one job. For example:

- Zymase can only convert glucose to ethanol and carbon dioxide.
- amylase can only break down starch to maltose.

Enzymes work best under strict conditions of temperature and pH. Look at the graphs below, which show how the activity of amylase changes with pH and temperature.

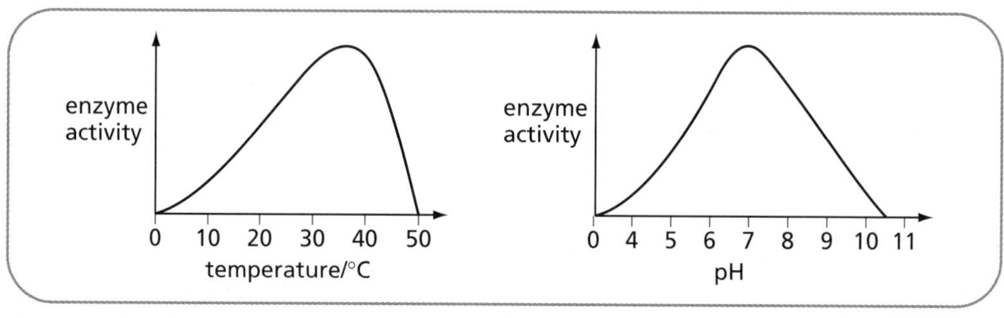

Figure 15.11 The effect of temperature and pH on amylase activity

The graphs show that the **optimum** conditions for amylase are a temperature of 37 °C and a pH of 7. This is the same as the conditions found in your mouth.

The alkanol series

Ethanol, the alcohol produced by fermentation, is a member of a series of similar compounds called the **alkanol** series. This is a **homologous series**, similar to the alkanes, except they contain an -OH group in place of one of the hydrogen atoms. The first six members of the alkanol series are shown in the table below.

Alkanol	Molecular formula	Boiling point (°C)	Full structural formula
methanol	CH_3OH	65	H–C(H)(H)–O–H
ethanol	C_2H_5OH	79	H–C(H)(H)–C(H)(H)–O–H
propanol	C_3H_7OH	97	H–C(H)(H)–C(H)(H)–C(H)(H)–O–H
butanol	C_4H_9OH	117	H–C(H)(H)–C(H)(H)–C(H)(H)–C(H)(H)–O–H
pentanol	$C_5H_{11}OH$	138	H–C(H)(H)–C(H)(H)–C(H)(H)–C(H)(H)–C(H)(H)–O–H
hexanol	$C_6H_{13}OH$	158	H–C(H)(H)–C(H)(H)–C(H)(H)–C(H)(H)–C(H)(H)–C(H)(H)–O–H

Like all homologous series they have similar chemical properties, fit a general formula and show a gradual change in physical properties down the series. For example, the boiling points of the alkanols increase as their molecular size increases.

Key Points

General formula of alkanols is $C_n H_{2n+1} OH$

Summary

- Carbohydrates, the energy foods, contain carbon, hydrogen and oxygen only.
- Three types: monosaccharides, $C_6H_{12}O_6$ glucose; disaccharides, $C_{12}H_{22}O_{11}$ sucrose; polysaccharides, $(C_6H_{10}O_5)_n$ starch.
- Benedict's solution changes blue to orange with glucose, fructose and maltose.
- Iodine solution changes brown to black with starch.
- Photosynthesis in green plants takes in carbon dioxide and water, traps sun's energy with chlorophyll and forms glucose and oxygen.
- Respiration in plants and animals uses glucose and oxygen to release energy and forms carbon dioxide and water.
- Combustion of carbohydrates is the same overall process as respiration.
- The carbon cycle involves photosynthesis, respiration and combustion.
- Carbon dioxide levels are rising, as there is more combustion and less forest.
- Carbon dioxide causes the greenhouse effect and this may lead to global warming.
- In condensation reactions, water is eliminated to join molecules, e.g. glucose → starch.
- In hydrolysis reactions, water is added to split up molecules, e.g. starch → glucose.
- Hydrolysis occurs during digestion, where acids or enzymes break down starch.
- Enzymes are biological catalysts that work best at a particular temperature and pH.
- The alcohol in alcoholic drinks is ethanol (C_2H_5OH).
- Ethanol is made by the fermentation of glucose using the enzyme zymase in yeast.
- Distillation is used to increase the alcohol content of spirits.
- Ethanol is a member of the alkanol series. General formula: $C_nH_{2n+1}OH$.